LOEB CLASSICAL LIBRARY

FOUNDED BY JAMES LOEB 1911

HIPPOCRATES

II

LCL 148

HIPPOCRATES

VOLUME II

EDITED AND TRANSLATED BY

PAUL POTTER

HARVARD UNIVERSITY PRESS
CAMBRIDGE, MASSACHUSETTS
LONDON, ENGLAND
2023

First published 2023

Library of Congress Control Number 2023001177
CIP data available from the Library of Congress

ISBN 978-0-674-99758-5

*Composed in ZephGreek and ZephText by
Technologies 'N Typography, Merrimac, Massachusetts.
Printed on acid-free paper and bound by
Maple Press, York, Pennsylvania*

CONTENTS

LIST OF HIPPOCRATIC WORKS SHOWING THEIR DIVISION INTO VOLUMES IN THIS EDITION

GENERAL INTRODUCTION

W. H. S. Jones' second Loeb *Hippocrates* volume (1923), which the present volume replaces,[1] is based throughout on complete collations of all the then-identified independent Greek witnesses. *Prognostic* and *Regimen in Acute Diseases* derive from volume 1 of H. Kühlewein's *Hippocratis Opera* (Leipzig, 1894), *The Art* from Gomperz' monograph *Apologie der Heilkunst* (Vienna, 1890; rev. ed. Leipzig, 1910), *Breaths* from Nelson's dissertation *Die hippokratische Schrift* ΠΕΡΙ ΦΥΣΩΝ (Uppsala, 1909), and the remaining *Sacred Disease*, *Law*, *Decorum*, and *Dentition* from Jones' own collations made from photographs. In the intervening century, almost all of these texts have benefited to a greater or less extent from further study, including the appearance of new critical editions and commentaries.

The eight Hippocratic writings in this volume vary widely in both their presentation and their purpose. The first two treatises are practical manuals instructing physicians respectively on how to predict the course and outcome of acute diseases from their clinical signs, and how to apply

[1] Omitted from this volume is the first chapter of *Physician*, since the entire treatise is now present at Loeb *Hippocrates* vol. 8 (LCL 482), pp. 291–311.

appropriate dietetic measures for them, in particular barley gruel, drinks, and baths. Both individually, and later as components of the Salernitan medical curriculum embodied in the *Articella*, *Prognostic* and *Regimen in Acute Diseases* enjoyed a wide diffusion, being—after *Aphorisms*—for many centuries the most often copied and translated Hippocratic works.

The next three works are monographs, each presenting arguments in favor of some specific hypothesis; all make skillful use of the rhetorical art in their clear exposition, their polished style, and the techniques of persuasion they employ. *Sacred Disease* argues that the name "sacred disease" is a misnomer, because all diseases are at the same time understandable both in terms of nature—anatomy and physiology—and in terms of such divine forces as the winds, the seasons, and the celestial bodies. The author disparages charlatans who claim to be able to cure this disease by magical and religious procedures, and emphasizes the predominant role of the brain in human mental and neurological functions. *The Art* defends the existence of the medical art against those who call it into question and detract from its credibility; in fact, the treatise displays such a proficiency in the use of rhetoric that Th. Gomperz attributed it to a sophist rather than a physician, although this viewpoint has not received general acceptance. *Breaths* is an admirably presented discussion of the wide role "air" in its various forms plays in life and health, laying the foundation for a "pneumatic" medicine that later in antiquity became an actual school.

Of the remaining three texts, *Law*, sketching a new model of medical education called for by medicine's current lack of repute, and *Decorum*, summarizing a public

address on medical wisdom—as an intellectual achievement and a collection of attitudes and behaviors—are both shrouded in a degree of obscurity, which scholars have attempted to explain as being due either to the esoteric milieu to which they belonged or to some external factor that prevented their final completion. The short tract *Dentition* consists of a loose collection of pediatric aphorisms dealing mainly with the nursing of infants and ulcerations of their tonsils, uvula, and throat.

HISTORY AND CONSTITUTION OF THE TEXT

C′	Parisinus Graecus Supplement 446 X c.
M	Marcianus Venetus Graecus 269 X c.
Θ	Vindobonensis Medicus Graecus IV X/XI c.
A	Parisinus Graecus 2253 XI c.
V	Vaticanus Graecus 276 XI/XII c.
Urb	Urbinas Graecus 64 XII c.
I	Parisinus Graecus 2140 XIII/XIV c.
H	Parisinus Graecus 2142
Ha	older part[2] XII/XIII c.
Hb	newer part XIV c.
Amb[a]	Ambrosianus Graecus 134 (B 113 sup.) XIII/XIV c.
R	Vaticanus Graecus 277 XIV c.
Recentiores	approximately twenty manuscripts XV/XVI c.

[2] Folios 46, 49, 55–78, and 80–308.

The *stemma codicum* appearing as Figure 1 provides an overview of the interdependencies among the manuscripts containing the eight treatises in this volume. The particular treatises are transmitted in the following independent witnesses:[3]

Prognostic	C′ M V
Regimen in Acute Diseases	M A V
Sacred Disease	M Θ
The Art	M A Urb
Breaths	M A Urb
Law	M V Amb[a]
Decorum	M
Dentition	V

NOTE ON TECHNICAL TERMS[4]

The following expressions require clarification as they are employed in this volume with particular significance.

ἱερός, θεός, τὸ θεῖον/sacred, god, the divine: these terms, which were often used interchangeably at the time of the Hippocratic Collection, denote the property of having a meaning, cause, or force, understandable by, or in relationship to, the gods, a realm only partly comprehensible to human beings. Cf. Heraclitus B 93 in DK: "The lord to

[3] For additional witnesses, see below, pp. 5f., 138, 189, and 223.

[4] For the explanation of other technical terms occurring in the present volume, see also Loeb *Hippocrates* vol. 1 (LCL 147), pp. lxxiii–lxxv; vol. 8 (LCL 482), p. 10f.; vol. 10 (LCL 520), pp. x–xiii; and vol. 11 (LCL 538), p. xvi f.

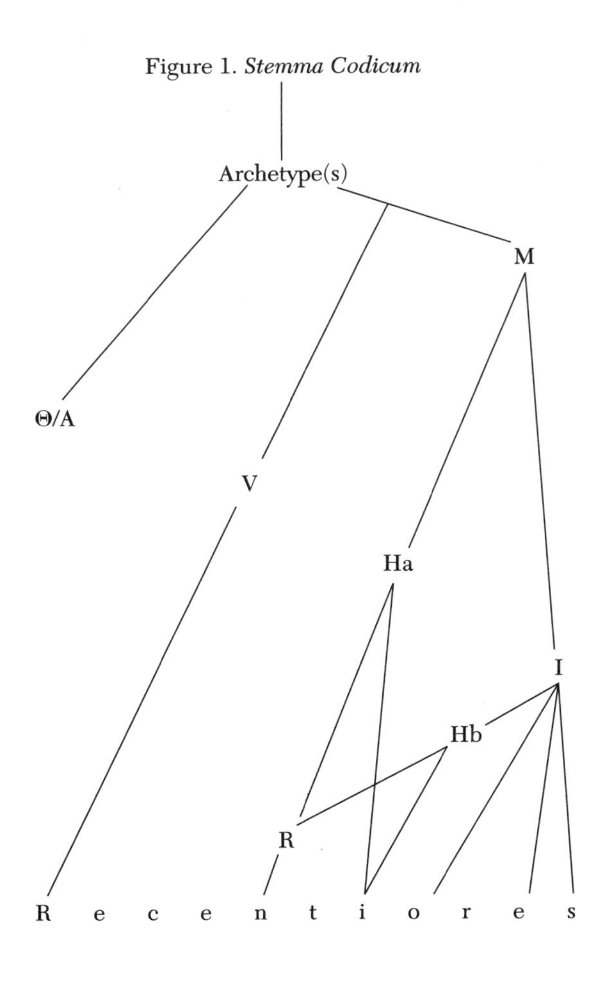

Figure 1. *Stemma Codicum*

whom the oracle in Delphi belongs neither tells nor conceals, but gives a sign (σημαίνει)." In a medical context, "divinity" is often attributed to natural phenomena, a qualification found frequently in nonmedical literature as well, e.g. Aeschylus, *Prometheus Bound*, where Prometheus prays (trans. Ph. Vellacott):

> O divinity of the ether, and swift-winged winds, and springs of rivers,
> O unnumbered laughter of the waves of the sea,
> O Earth, mother of all things,
> On you, and on the all-seeing circle of the sun, I call.[5]

ἴχωρ/serum: this word is used in health for the watery part of bodily fluids, e.g., the serum of the blood, in trauma for the sanies drained from a lesion, and in diseases for a fluid that has become unhealthy: e.g., "Ardent fever occurs when the small vessels are dried up in summer, and attract sharp bilious sera" (*Regimen in Acute Diseases* [*Appendix*] ch. 1).[6]

σοφία/wisdom: this Greek term has the same ambiguity of meaning as the English word "wisdom," which the *OED* defines as:

> 1. Capacity of judging rightly in matters relating to life and conduct; soundness of judgement in the choice of means and ends; sometimes, less strictly,

[5] For a study of the "divine" in *Sacred Disease*, see Philip J. van der Eijk, "The 'theology' of the Hippocratic Treatise *On the Sacred Disease*," in Eijk, pp. 45–73.

[6] Loeb *Hippocrates* vol. 6 (LCL 473), p. 229.

> sound sense, esp. in practical affairs. . . . 2. Knowledge (esp. of a high or abstruse kind); enlightenment, learning, erudition, in early use often = philosophy, science. Also, practical knowledge or understanding, expertness in an art.

In different passages of *Decorum* these two senses are understood with varying degrees of emphasis: e.g., ch. 5, where the moral aspect of "wisdom" is stressed; and ch. 1, where "wisdom" is synonymous with an art, in particular medicine.

REFERENCES[7]

EDITIONS, TRANSLATIONS, AND COMMENTARIES

Hippocratic Collection

Early Works

Calvus	*Hippocratis Coi . . . octoginta volumina . . . per M. Fabium Calvum, Rhavennatem . . . latinitate donata . . .* Rome, 1525.
Aldina	*Omnia opera Hippocratis . . . in aedibus Aldi & Andreae Asulani soceri.* Venice, 1526.
Froben	*Hippocratis Coi . . . libri omnes . . . [per Ianum Cornarium].* Basel, 1538.
Cornarius	*Hippocratis Coi . . . opera . . . per Ianum Cornarium . . . Latina lingua conscripta.* Venice, 1546.
Cornarius in marg.	Marginal notes by Janus Cornarius in a copy of the Aldine edition presently kept in the Göttingen University Library.

[7] Supplementary bibliographical information is given in the introductions to the particular treatises.

Zwinger	*Hippocratis Coi . . . viginti duo commentarii . . . Theod. Zvingeri studio & conatu.* Basel, 1579.
Foes	*Magni Hippocratis . . . opera quae extant . . . Latina interpretatione & annotationibus illustrata, Anutio Foesio . . . authore.* Geneva, 1657–1662.
Linden	*Magni Hippocratis Coi opera omnia graece & latine edita . . . industria & diligentia Joan. A. Vander Linden.* Leiden, 1665.
Clifton	Clifton, F. *Hippocrates upon . . . Prognosticks, in Acute Cases Especially.* London, 1734.
Mack	*Hippocratis opera omnia . . . studio et opera Stephani Mackii.* Vienna, 1743–1749.
Grimm	Grimm, J. F. K. *Hippokrates Werke aus dem griechischen. . . .* Altenburg, 1781–1792.

Post-Eighteenth Century

Adams	Adams, F. *The Genuine Works of Hippocrates.* London, 1849.
Alexanderson	Alexanderson, B. *Die hippokratische Schrift Prognostikon.* Gothenburg, 1963.
Alex *TK*	———. *Textkritischer Kommentar zum hippokratischen Prognostikon. . . .* Gothenburg, 1968.
Chadwick	Chadwick, J., and W. N. Mann. *The Medical Works of Hippocrates.* Oxford, 1950.

Daremberg	Daremberg, Ch. *Oeuvres choisies d' Hippocrate*^{2}. Paris, 1855.
Dietz	Dietz, Fr. *Hippocratis De morbo sacro*. Leipzig, 1827.
Diller *Schr.*	Diller, H. *Hippokrates. Schriften . . . übersetzt*. Hamburg, 1962.
Ermerins	Ermerins, F. Z. *Hippocratis . . . reliquiae*. Utrecht, 1859–1864.
Fleischer	Fleischer, U. *Untersuchungen zu den pseudohippokratischen Schriften* Παραγγελίαι, Περὶ ἰητροῦ, *und* Περὶ εὐσχημοσύνης. Berlin, 1939.
Fuchs	Fuchs, R. *Hippokrates, sämmtliche Werke. Ins Deutsche übersetzt. . . .* Munich, 1895–1900.
Gomperz	Gomperz, Th. *Die Apologie der Heilkunst*^{2}. Leipzig, 1910.
Heiberg	Heiberg, J. *Hippocratis Opera*. CMG I 1. Leipzig and Berlin, 1927.
Jones	Jones W. H. S. *Hippocrates with an English Translation*. London and New York, 1923–1931.
Kühlewein	Kühlewein, H. *Hippocratis Opera*. Leipzig, 1894–1902.
Littré	Littré, E. *Oeuvres complètes d'Hippocrate*. Paris, 1839–1861.
Nelson	Nelson, A. *Die hippokratische Schrift* ΠΕΡΙ ΦΥΣΩΝ. Uppsala, 1909.
Reinhold	Reinhold, C. H. Th. *Ἱπποκράτης/Ψευδωνύμως Ἱπποκράτεια*. Athens, 1864–1867.

REFERENCES

Other Authors

Aretaeus	Hude, C. *Aretaeus*[2]. CMG II. Berlin, 1958.
Boudon	Boudon-Millot, V. *Galien . . . Sur ses propres livres*. Budé I. Paris, 2007.
Caus. Aff.	Helmreich, G. *Handschriftliche Studien zu Galen. II. Teil*. Progr. Ansbach, 1911.
DK	Diels, H., and W. Kranz. *Die Fragmente der Vorsokratiker*[6]. Berlin, 1951.
Duffy	Duffy, J. *Stephanus the Philosopher. A Commentary on the Prognosticon of Hippocrates*. CMG XI 1,2. Berlin, 1983.
Erotian	Nachmanson, E. *Erotiani Vocum Hippocraticarum collectio*. Gothenburg, 1918.
Galen	Kühn, C. G. *Claudii Galeni opera omnia* . . . Leipzig, 1825–1833.
Heeg	Heeg, J. *Galeni In Hippocratis Prognosticum Comm. III*. CMG V 9,2. Leipzig and Berlin, 1915, pp. xviii–xxix and 195–378.
Perilli	Perilli, L. *Galeni Vocum Hippocratis Glossarium*. CMG V 13,1. Berlin, 2017.
Vellacott	Vellacott, Ph. *Aeschylus. Prometheus Bound*. . . . Harmondsworth, 1961.
Westerink	Westerink, L. G. *Stephanus of Athens. Commentary on Hippocrates' Aphorisms*. CMG XI 1,3,1–3. Berlin 1985–1995.

REFERENCES

GENERAL WORKS

Andorlini	Andorlini, I., ed. *Greek Medical Papyri I/II*. Florence, 2001/2009.
Budé	Collection des universités de France publiée sous le patronage de l'Association Guillaume Budé.
CMG	Corpus Medicorum Graecorum
Diels	Diels, H. "Hippokratische Forschung IV." In *Hermes* 48 (1913): 378–407.
Diller *KSAM*	Diller, H. *Kleine Schriften zur antiken Medizin*. Berlin, 1973.
Dönt	Dönt, H. *Die Terminologie von Geschwür, Geschwulst und Anschwellung im Corpus Hippocraticum.* Diss., Vienna, 1968.
Ducatillon	Ducatillon, J. *Polémiques dans la Collection Hippocratique*. Paris, 1977.
Durling	Durling, R. J. *A Catalogue of Sixteenth Century Printed Books in the National Library of Medicine.* Bethesda, 1967.
Eijk	van der Eijk, Ph. J. *Medicine and Philosophy in Classical Antiquity.* Cambridge, 2005.
Fischer	Fischer, K.-D. "Praenostica. Die rezeption des Prognostikons im Frühmittlelalter." In *La science médicale antique. Nouveaux regards*, edited by V. Boudon-Millot et al., pp. 189–247. Paris, 2007.

García-Bal.	García-Ballester, L. "The *New Galen*: A Challenge to Latin Galenism in Thirteenth-Century Montpellier." In *Text and Tradition. Studies in Ancient Medicine and Its Transmission*, edited by K.-D. Fischer et al. Leiden, 1998.
Grensemann	Grensemann, H. *Knidische Medizin I/II*. Berlin, 1975; *Hermes Einzelschriften* 51. Stuttgart, 1987.
Heidel	Heidel, W. A. "Hippocratea I." *Harvard Studies in Classical Philology* 25 (1914): 139–205.
Hind et al.	Hind, M., and D. Leith. *The Oxyrhynchi Papyri* LXXX. London, 2014–2015.
Index Hipp.	Kühn, J.-H., U. Fleischer et al. *Index Hippocraticus*. Göttingen, 1986–1989; Anastassiou, A., and D. Irmer, *Supplement*. 1999.
Jouanna	Jouanna, J. *Hippocrates*. Translated by M. B. DeBevoise. Baltimore, 1999.
Jouanna *Arch.*	———. *Hippocrate. Pour une archéologie de l' école de Cnide*. Paris, 1974.
Kibre	Kibre, P. *Hippocrates Latinus*. Rev. ed. New York, 1985.
Klamroth	Klamroth, M. "Über die Auszüge aus griechischen Schriftstellern bei al-Jaʿqûbî." *Zeitschrift der deutschen morgenländischen Gesellschaft* 40 (1886): 204–33.

Kristeller	Kristeller, P. O. "Bartholomaeus, Musandinus and Maurus of Salerno and Other Early Commentators of the *Articella.*" *Italia medioevale e umanistica* 19 (1976): 57–87.
Lonie *Cos*	Lonie, I. M. "Cos versus Cnidus and the Historians." *History of Science* 16 (1978): 42–92.
LSJ	Liddell, H. G., R. Scott, and H. S. Jones, eds. *A Greek-English Lexicon*[9]. Oxford, 1940.
Manetti/Ros.	Manetti, D., and A. Roselli. "Galeno commentatore di Ippocrate." In *Aufstieg und Niedergang der römischen Welt* (*ANRW*) II 37,2, edited by W. Haase and H. Temporini, pp. 1529–635. Berlin, 1994.
Marganne	Marganne, M.-H. *Inventaire analytique des papyrus grecs de médecine.* Geneva, 1981.
Mondrain	Mondrain, B. "Lire et copier Hippocrate . . . au XIVe siècle," In *Ecdotica e ricezione dei testi medici greci*, edited by V. Boudon-Millot et al., pp. 359–410. Naples, 2006.
OED	Murray, J. A. H. et al., eds. *The Oxford English Dictionary*. Oxford, 1933; *Supplement*, 1972–1986.
Nachmanson	Nachmanson, E. *Erotianstudien.* Uppsala, 1917.

Roselli	Roselli, A. "Citazioni Ippocratiche in Demetrio Lacone." *Cronache Ercolanesi* 18 (1988): 53–57.
Schullian	Schullian, D., and F. Sommer. *A Catalogue of Incunabula and Manuscripts in the Army Medical Library.* New York, 1950.
Sezgin	Sezgin, F. *Geschichte des arabischen Schriftums. Band III. Medizin – Pharmazie – Zoologie – Tierheilkunde bis ca 430 H.* Leiden, 1970.
Skoda	Skoda, F. *Médecine ancienne et métaphore.* Paris, 1988.
Smith *Coans*	Smith, W. D. "Galen on Coans and Cnidians." *Bulletin of the History of Medicine* 47 (1973): 569–85.
Testimonien	Anastassiou, A., and D. Irmer. *Testimonien zum Corpus Hippocraticum.* Göttingen, (vol. I) 2006; (vol. II,1) 1997; (vol. II,2) 2001; (vol. III) 2012.
Thivel	Thivel, A. *Cos et Cnide?* Nice-Paris, 1981.
Ullmann	Ullmann, M. *Die Medizin im Islam.* Leiden, 1970.
Wilamowitz	Wilamowitz-Möllendorf, U. von. "Die hippokratische Schrift *περὶ ἱρῆς νούσου.*" *Sitzungsbericht der k. p. Akademie d. Wissenschaften zu Berlin,* 1901: 2–23. = Wilamowitz. *Kleine Schriften* III: Berlin, 1969: 278–302.

GENERAL BIBLIOGRAPHY

EDITIONS, TRANSLATIONS AND COMMENTARIES

Bourbon, F. *Hippocrate. Femmes stériles, Maladies des jeunes filles, Superfétation, Excision du fœtus*. Budé XII (4). Paris, 2017.

Ecca, G. *Die hippokratische Schrift Praecepta*. Wiesbaden, 2016.

García Gual, C. et al. *Tratados hipocráticos . . . introducciones, traducciones y notas*. Madrid, 1983–2003.

Giorgianni, F. *Hippokrates. Über die Natur des Kindes* (De genitura *und* De natura pueri). Wiesbaden, 2006.

Hanson, M. *Hippocrates. On Head Wounds*. CMG I 4,1. Berlin, 1999.

Jouanna, J. *Hippocrate. Maladies II*. Budé X (2). Paris, 1983.

———. *Hippocrate. Des Vents. De l'Art*. Budé V (1). Paris, 1988.

———. *Hippocrate. De la maladie sacrée*. Budé II (3). Paris, 2003.

———. *Hippocrate. Pronostic*. Budé III (1). Paris, 2013.

Overwien, O. *Hippokrates. Über die Säfte*. CMG I 3,1. Berlin, 2014.

Roselli, A. *Ippocrate. La malattia sacra*. Venice, 1996.

GENERAL WORKS

Bliquez, L. *The Tools of Asclepius*. Leiden, 2014.

Bruno Celli, B. *Bibliografía Hipocrática*. Caracas, 1984.

Byl, S. "Les dix dernières années (1982–1992) de la recherche hippocratique." In *Université Jean Monnet-Saint Etienne. Centre Jean Palerne. Lettre d'Informations* 22 (1993): 1–39.

Craik, E. M. *The 'Hippocratic' Corpus: Content and Context*. London, 2015.

Dean-Jones, L. A., and R. M. Rosen, eds. *Ancient Concepts of the Hippocratic. Papers Presented at the XIIIth International Hippocratic Colloquium, Austin, Texas 2008*. Leiden, 2015.

Eijk, Ph. J. van der, ed. *Hippocrates in Context. Papers Read at the XIth International Hippocratic Colloquium (Newcastle-upon-Tyne 2002)*. Leiden, 2005.

Eijk, Ph. J. van der, H. F. Horstmanshoff, and P. H. Schrijvers, eds. *Ancient Medicine in its Socio-cultural Context*. Amsterdam, 1995.

Fichtner, G. *Corpus Hippocraticum. Verzeichnis der hippokratischen und pseudohippokratischen Schriften*. Berlin, 2015.

Flashar, H. *Hippokrates. Meister der Heilkunst*. Munich, 2016.

Harris, W. V., ed. *Mental Disorders in the Classical World*. Leiden, 2013.

Horstmanshoff, H. F., ed. *Hippocrates and Medical Education: Selected Papers Read at the XII International Hippocrates Colloquium*. Leiden, 2010.

Jouanna, J. *Greek Medicine from Hippocrates to Galen*. Leiden, 2012.

Leven, K.-H., ed. *Antike Medizin. Ein Lexikon.* Munich, 2005.

Maloney, G. *Index Inverses du vocabulaire Hippocratique.* Québec, 1987.

Maloney, G., and W. Frohn. *Concordance des oeuvres Hippocratiques.* Québec, 1984 and Hildesheim, 1986.

Maloney, G., and R. Savoie. *Cinq cents ans de bibliographie Hippocratique.* Québec, 1982.

Nutton, V. *Ancient Medicine.* London, 2004.

Oser-Grote, C. M. *Aristoteles und das Corpus Hippocraticum. Die Anatomie und Physiologie des Menschen.* Stuttgart, 2004.

Perilli, L., Ch. Brockmann, K.-D. Fischer, and A. Roselli, eds. *Officina Hippocratica.* Berlin, 2011.

Pormann, P., ed. *Cambridge Companion to Hippocrates.* Cambridge, 2018.

Thivel, A., and A. Zucker, eds. *Le normal et le pathologique dans la Collection hippocratique. Actes du Xème colloque international hippocratique (1999).* Nice, 2001.

PROGNOSTIC

INTRODUCTION

Prognostic is a comprehensive, practical manual explaining how the physician can, by observing and interpreting various clinical signs, deduce the course a patient's acute disease has, and will run, in particular whether it will end fatally or in recovery. The author, in spite of his obvious familiarity with generally shared Hippocratic pathological and nosological ideas, takes the exceptional position that such signs have fixed prognostic significance independent of season, geographical location, and disease diagnosis.[1] The account is organized by topic in the following chapters:

1 Importance of prognosis
2 Signs in the face
3 Posture of the body
4 Movement of the hands
5 Respiration
6 Sweating
7 Signs in the abdomen

[1] The subject of prognosis also plays a notable role in the aphoristic works *Prorrhetic I*, *Coan Prenotions*, and *Aphorisms*; the pathological treatises *Diseases III*, *Internal Affections*, and *Diseases of Women I*; and the monograph *Prorrhetic II*.

8 Edemas
9 Temperature; heaviness
10 Sleep and waking
11 Stools
12 Urines
13 Vomiting
14/15 Sputum
16/17 Suppuration in the chest
18 Abscessions after pneumonia
19 Pains together with fever
20 Critical days
21–23 Pains in the head and throat
24 Fevers
25 Summation

There is evidence that *Prognostic* was already part of the incipient Hippocratic Collection in the third century BC: an entry (A 5) in Erotian's glossary on the word ἀλλοφάσσοντες (in *Prognostic* ch. 20) alludes to the opinions of the two early lexicographers Xenocritus of Cos (fl. ca. 300) and Bacchius of Tanagra (ca. 275–200) on its meaning; Bacchius also includes entries on words drawn from the text of *Prognostic* in *syntaxes* 1 and 2 of his *Words*.[2]

Erotian in the first century AD lists *Prognostic* in his Hippocratic census under the semeiotic works and comments on about thirty terms from the treatise among his glosses, citing Bacchius in five of these.[3] Galen of Pergamum (129–ca. 210) refers to *Prognostic* frequently

[2] See Loeb *Hippocrates* vol. 1 (LCL 147), pp. xlvii–xlix.
[3] Cf. Nachmanson, pp. 268–73.

throughout his works,[4] includes in his *Explanation of Difficult Words in Hippocrates* three glosses certainly (α66, μ15, and ρ15) and two possibly (α70, δ5) from the treatise,[5] and has left an extensive commentary on the text.[6] A sixth-century commentary on *Prognostic* by Stephan the Philosopher is also preserved.[7] Among ancient writers, many references and citations from *Prognostic* are to be found.[8]

The great attention *Prognostic* enjoyed in antiquity extended into the middle ages, as is evidenced by the relatively large number of manuscript copies, translations, commentaries, and quotations that have survived: these sources contribute to our knowledge of both the original text and its subsequent reception. The following simplified account of textual sources is organized in increasing distance from the Greek text, beginning in antiquity.

1. The Greek text preserved in ancient papyri:

Π_{20} P. Herculaneum 831 (150–50 BC): a citation from *Prognostic* ch. 7.[9]

Π_{23} P. Tebtunis II 678 (I/II c.): texts from chs. 14–15 of *Prognostic*.[10]

Π_{21} P. Oxyrhynchus 5223 (II c.): a passage from *Prognostic* ch. 7.[11]

[4] Cf. *Testimonien* vol. II,1, pp. 394–432; vol. II,2, pp. 302–44.
[5] See the *Prognostic* passages listed by Perilli, p. 416.
[6] See Heeg, pp. xviii–xxix and 195–378.
[7] See Duffy.
[8] Cf. *Testimonien* vol. I, pp. 400–425.
[9] See Roselli, p. 55.
[10] See Andorlini, vol. 2, pp. 15–33.
[11] See Hind et al., pp. 11–15.

Π_7 P. Antionoopolis I 28 (III/IV c.): a passage from *Prognostic* chs. 24–25.[12]

Π_{22} P. Oxyrhynchus inv. 3 1B.82/c(1)b (VI c.): the first seven lines of *Prognostic*.[13]

2. Medieval and renaissance Greek manuscripts of the Hippocratic tradition:

About sixty Greek manuscripts are known that contain either all (approx. fifty) or part (approx. ten) of *Prognostic*. Roughly half of these are the same manuscripts recorded for the Collection in general in Loeb *Hippocrates* volume 1: they include the primary witnesses M and V, and the occasional independent witnesses C′ and Urb (fols. 96^r–102^r).[14] The remaining thirty manuscripts are mostly miscellaneous medical manuscripts containing, together with one or more Hippocratic texts, various medical writings by other ancient and Byzantine authors. Alexanderson demonstrated that in most of these miscellaneous manuscripts the *Prognostic* text is taken over from extant members of the M and V families.[15] But to the two occasional independent witnesses C′ and Urb noted above, Alexanderson added five more:

G′ Laurentianus Graecus plut. 75,3 (X/XI c.)
S Laurentianus Graecus plut. 74,11 (XIII c.)
Y Vaticanus Palatinus Graecus 199 (XIII/XIV c.)

[12] See Marganne, p. 72.

[13] See Jouanna *Prog.*, p. cxxiii.

[14] See Loeb *Hippocrates* vol. 1 (LCL 147), p. lxiv; Alexanderson, pp. 69–124 and 251; Alex *TK*, pp. 12–16; Jouanna *Prog.*, pp. lxxxiii–cxxii.

[15] Cf. Alexanderson, pp. 100–120.

N Parisinus Graecus 2316 (XV c.)
Z Vaticanus Graecus 2254 (excerpts) (X/XI c.)[16]

This group of seven manuscripts, which share certain readings only among themselves, must be in some way affiliated, but no attempt to establish a clear stemmatic ordering of their interdependencies has so far succeeded, leading Jouanna to suggest calling them "the constellation of manuscript C′," and assessing their readings on a case-by-case basis.[17]

3. Medieval and renaissance Greek manuscripts of Galen's *Commentary to Prognostic*:

Heeg, J. *Galeni In Hippocratis Prognosticum Comm. III*. CMG V 9,2. Leipzig and Berlin, 1915, pp. xviii–xxix and 195–378.

In this edition, Heeg presents a *stemma codicum* containing nineteen Greek manuscripts, of which he identifies four as being independent. The commentary consists of succeeding sections, each citing a one/two sentence passage from the Hippocratic text (lemma), followed by Galen's comments on it. Besides these complete *Commentary* manuscripts, Alexanderson identifies a dozen other manuscripts containing the Greek text of *Prognostic* alone (without Galen's comments), which has been created by extracting and reassembling the lemmata of the *Commentary*.[18]

[16] Cf. Alexanderson, pp. 92–100.
[17] Cf. Jouanna *Prog*., pp. cii–cxxii.
[18] Cf. Alexanderson, pp. 120–24; Jouanna *Prog.*, cliv–clxii.

4. Late ancient or early medieval Latin translations from now lost Greek manuscripts:

Lat1 (inc. *Medicum existimo perfectum esse praescientiam affectantem*), which is preserved in two manuscripts

A Ambrosianus Latinus G 108 (IX/X c.)
Q Monacensis Latinus 11343 (XIII c.)

and edited by Alexanderson.[19]

Lat2 (inc. *Medicum videtur mihi optimum esse providentiam imitantem*), which contains only *Prognostic* chs. 1–5, is contained in two ninth century manuscripts in the Saint Gallen Stiftsbibliothek (Switzerland), numbers 44 and 751.[20]

5. Arabic translations of Galen's *Commentary to Prognostic* made from now lost Greek manuscripts.[21]

Both the Hippocratic lemmata and Galen's comments on them were translated into Arabic by Ḥunain ibn Isḥāq (d. 873) and members of his school: passages from these translations are cited by several Arabic medical writers, and there are a number of Arabic commentaries on the texts. Ḥunain's Arabic translation of the Hippocratic text alone (without Galen's comments) was edited from three

[19] See Alexanderson, pp. 124–56.

[20] See Alexanderson, p. 155f.; Jouanna *Prog.*, pp. cxxxvii–cxli; Fischer.

[21] Cf. Ullmann, p. 29; Sezgin, p. 32f.; Alexanderson, pp. 156–73; Jouanna *Prog.*, pp. clxii–clxxvii.

of many manuscripts by Klamroth in 1886 and has recently been published in a German translation by O. Overwien and U. Vagelpohl.[22]

6. Latin translations from the Arabic translation of Galen's *Commentary to Prognostic*:[23]

Lat3 (inc. *Omnis qui medicine artis studio*) is sometimes attributed to the translator Constantinus Africanus (d. 1087).

Lat4 (inc. *Videtur mihi quod/ut ex melioribus rebus*) is sometimes attributed to the translator Gerard of Cremona (d. 1187).

Lat3 and Lat4 together are extant in over two hundred and fifty manuscripts dating from the twelfth to the fifteenth century. In some copies, only the Hippocratic text is contained, while in others Galen's commentary too is preserved.[24] The translations in these manuscripts often form part of the *Articella*, a standardized collection of medical works, apparently of Salernitan origin, that in its earliest state consisted of Latin versions of Johannitius (= Ḥunain), *Introduction to Galen's Microtechne*; Philaretus, *On Pulses*; Theophilus, *On Urines*; Hippocrates, *Aphorisms*; Hippocrates, *Prognostic*.[25]

7. Lat5 (inc. *Medicum videtur mihi optimum esse providentiam adinvenire*), derived from a Greek manuscript of

[22] See Jouanna *Prog.*, pp. 273–98, where the German translation is published.

[23] See Alexanderson, pp. 170–72; Jouanna, pp. clxxviii–clxxxiv.

[24] See Kibre, pp. 199–221.

[25] See Kristeller, pp. 59f., 65–68, 75–81.

Galen's *Commentary to Prognostic* earlier than any of those extant, is preserved in Neopolitanus Latinus VIII D 25 (dated 1380); this translation has been attributed by S. Fortuna to Niccolò da Reggio.[26]

Prognostic was first printed in Latin translation in an undated edition of the *Articella* before 1479, and in the sixteenth century was often printed in a new Latin translation by W. Kopp that first appeared in 1511 (Durling, 2440). The treatise is present in all the collected editions and translations, including Mack, Grimm, Adams, Daremberg, Kühlewein, Chadwick, and Diller *Schr*. Littré (vol. 2, pp. 103–9) records the special studies devoted to *Prognostic* from the sixteenth to the nineteenth century, and Jones (vol. 2, p. 5) draws attention to English translations by Peter Lowe (1597) and J. Moffat (1788), and to F. Clifton's collection of excerpts in his "Prognosticks, in Acute Cases Especially" (1734), pp. 250–389.

The relative richness of *Prognostic's* Greek transmission,[27] due in part to the existence of Galen's commentary, began to be recognized in the nineteenth century: Littré introduced C′ into his admittedly rather diffuse apparatus, while Kühlewein (followed by Jones) made it his primary source for the text; in addition, Kühlewein drew attention to the importance of Z, G′, and S, although he made only sparing use of their readings.[28] Two subsequent editors make much fuller use of the whole *Prognostic* tradition:

[26] See Jouanna *Prog*., pp. clxxxiv–clxxxvii, esp. n. 216.

[27] Cf. Mondrain, p. 361f.

[28] See Littré, vol. 2, p. 103, n. 2; Kühlewein, vol. 1, pp. 72–77.

Alexanderson, B. *Die hippokratische Schrift Prognostikon*. Gothenburg, 1963. (= Alexanderson); idem. *Textkritischer Kommentar zum hippokratischen Prognostikon*. Gothenburg, 1968. (= Alex *TK*)

Jouanna, J. *Hippocrate. Pronostic.* Budé III (1). Paris, 2013. (= Jouanna *Prog.*)

Alexanderson added Urb, Y, and N to the list of independent Greek witnesses and introduced all seven *sigla* into his apparatus. He also analyzes Galen's readings in both his *Commentary* and his other works, clarifies the origins of the Latin translations and prints the text of Lat1 in its entirety, and examines many readings from the Ḥunain Arabic translation on the basis both of Klamroth's edition and of the Arabic translation's translations into Latin.

Alexanderson sums up his views on the transmission of *Prognostic* with an assessment of the value of the different witnesses, based on a sample of twenty-eight important individual passages in which the correct text can be determined with relative confidence, sorting them according to the readings they each contain:

	Correct readings	Incorrect readings
C′	9	14
S	7	21
Y	7	21
Z	4	8
X	10	17
G′	10	16
N	12	14

	Correct readings	Incorrect readings
V	18	10
M	13	15
GalL(emmata)	12	14
GalT(estimonia)	18	2
Lat(1)	14	9
Ar(abic trans.)	17	8

These figures should lead to certain considerations: every manuscript—Greek, Latin, and Arabic—is full of mistakes; V, Galen, Lat, and Ar are particularly valuable; every passage must be evaluated on its own.[29]

Writing half a century later, Jouanna follows Alexanderson's suggestion, evaluating each passage independently on the basis of all its witnesses. In addition, he adds several new witnesses: for example, parallel texts in compilatory treatises of the Hippocratic collection, such as *Coan Prenotions*, *Crises*, and *Aphorisms*; a Syriac translation; a further Arabic translation of Galen's *Commentary*; a translation into German of Ḥunain's Arabic translation. Finally, he points out future steps to be taken in the study of the treatise.[30]

It will be clear how much the present Loeb edition is indebted to all these scholars.

[29] See Alexanderson, pp. 176–78.

[30] See Jouanna *Prog.*, pp. cclxxvi–cclxxviii.

ΠΡΟΓΝΩΣΤΙΚΟΝ

II 110 Littré

1. Τὸν ἰητρὸν δοκέει μοι ἄριστον εἶναι πρόνοιαν ἐπιτηδεύειν· προγινώσκων γὰρ καὶ προλέγων παρὰ τοῖσι νοσέουσι τά τε παρεόντα καὶ τὰ προγεγονότα καὶ τὰ μέλλοντα ἔσεσθαι, ὁκόσα τε παραλείπουσιν οἱ ἀσθενέοντες ἐκδιηγεύμενος, πιστεύοιτ' ἂν μᾶλλον γινώσκειν τὰ τῶν νοσεόντων πρήγματα, ὥστε τολμᾶν ἐπιτρέπειν τοὺς ἀνθρώπους σφέας ἑωυτοὺς τῷ ἰητρῷ. τὴν δὲ θεραπείην ἄριστ' ἂν ποιέοιτο προειδὼς τὰ ἐσόμενα ἐκ τῶν παρεόντων παθημάτων. ὑγιέας μὲν γὰρ ποιέειν ἅπαντας τοὺς ἀσθενέοντας[1] ἀδύνατον· τοῦτο γὰρ καὶ τοῦ προγινώσκειν τὰ μέλλοντα ἀποβήσεσθαι κρέσσον ἂν ἦν· ἐπειδὴ δὲ οἱ ἄνθρωποι ἀποθνῄσκουσιν, οἱ μὲν πρὶν ἢ καλέσαι τὸν ἰητρὸν ὑπὸ τῆς ἰσχύος τῆς νούσου, οἱ δὲ | καὶ ἐσκαλεσάμενοι παραχρῆμα ἐτελεύτησαν—οἱ μὲν ἡμέρην μίαν ζήσαντες, οἱ δὲ ὀλίγῳ πλείονα χρόνον—πρὶν ἢ τὸν ἰητρὸν τῇ τέχνῃ πρὸς

112

[1] ἀσθεν. MV: νοσέοντας C'

---

[1] This sentence has been construed differently by scholars on account of its ambiguous grammar. The word ἄριστον can be read either as a masculine accusative adjective modifying the

PROGNOSTIC

1. I consider the physician to be best who draws inferences in advance:[1] for if he knows beforehand and reports to patients the present, the past, and the future, and tells in detail what the ill have omitted to say, he will be the more trusted to understand their condition, so that they will have confidence in turning themselves over to their physician. Therapy is also best managed by someone who knows beforehand from the present sufferings what will happen later. Now to make all who are ill healthy is impossible—for that would be even better than predicting what was going to happen. Since however people do die, some before they call for the physician, due to the severity of their disease, and others who expire immediately after calling him in—some surviving for one day, and others for a little longer—before he can combat each disease with his art, it

noun ἰατρόν in the construction "the physician seems to me to be best . . ." or as a neuter form modifying the phrase "for a physician to practice forecasting." The early Latin translation Lat1 (perhaps sixth century) is an example of the former: "*Medicum existimo perfectum esse praescientiam affectantem*" (which suggests the Greek reading ἐπιτηδεύοντα for the manuscripts' ἐπιτηδεύειν); Jones, vol. 1, of the later: "I hold that it is an excellent thing for a physician to practice forecasting."

ἕκαστον νόσημα ἀνταγωνίσασθαι, γνόντα οὖν χρὴ τῶν παθῶν τῶν τοιούτων[2] τὰς φύσιας, ὁκόσον ὑπὲρ τὴν δύναμίν εἰσι τῶν σωμάτων, ἅμα δὲ καὶ εἴ τι θεῖον ἔνεστιν ἐν τῇσι νούσοισι, καὶ τούτων τὴν πρόνοιαν ἐκμανθάνειν. οὕτω γὰρ ἄν τις θαυμάζοιτό τε δικαίως καὶ ἰητρὸς ἀγαθὸς ἂν εἴη· καὶ γὰρ οὓς οἷόν τε περιγίνεσθαι ἔτι μᾶλλον ἂν δύναιτο διαφυλάσσειν ἐκ πλείονος χρόνου προβουλευόμενος πρὸς ἕκαστα, καὶ τοὺς ἀποθανουμένους τε καὶ σωθησομένους προγινώσκων τε καὶ προαγορεύων ἀναίτιος ἂν εἴη.

2. Σκέπτεσθαι δὲ χρὴ ὧδε ἐν τοῖσιν ὀξέσι νοσήμασι· πρῶτον μὲν τὸ πρόσωπον τοῦ νοσέοντος, εἰ ὅμοιόν ἐστι τοῖσι τῶν ὑγιαινόντων, μάλιστα δέ, εἰ αὐτὸ ἑωυτῷ· οὕτω γὰρ ἂν εἴη | ἄριστον, τὸ δὲ ἐναντιώτατον τοῦ ὁμοίου δεινότατον. εἴη δ' ἂν τὸ τοιόνδε· ῥὶς ὀξεῖα, ὀφθαλμοὶ κοῖλοι, κρόταφοι ξυμπεπτωκότες, ὦτα ψυχρὰ καὶ ξυνεσταλμένα καὶ οἱ λοβοὶ τῶν ὤτων ἀπεστραμμένοι καὶ τὸ δέρμα τὸ περὶ τὸ μέτωπον[3] σκληρόν τε καὶ περιτεταμένον καὶ καρφαλέον ἐόν· καὶ τὸ χρῶμα τοῦ ξύμπαντος προσώπου χλωρόν τε ἢ καὶ μέλαν ἐὸν καὶ πελιὸν ἢ μολιβδῶδες.[4] ἢν μὲν οὖν[5] ἐν ἀρχῇ τῆς νούσου τὸ πρόσωπον τοιοῦτον ᾖ καὶ μήπω οἷόν τε ᾖ τοῖσιν ἄλλοισι σημείοισι ξυντεκμαίρεσθαι, ἐπανερέσθαι χρή, μὴ ἠγρύπνηκεν ὁ ἄνθρωπος ἢ τὰ τῆς κοιλίης ἐξυγρασμένα ἦν[6] ἰσχυρῶς, ἢ λιμῶδές τι

114

[2] παθ. τ. τοιούτων MV: τῶν τοιούτων νοσημάτων C'
[3] μέτωπον MV: πρόσωπον C'

is thus necessary, by understanding the natures of such diseases, how they exceed the body's power, and at the same time whether there is anything divine about them, also to learn how to draw inferences about them in advance. For in this way a person will be justly admired, and in fact be a good physician. And furthermore he would be able both carefully to protect those who are capable of surviving, by contriving measures for them at an earlier time in each stage, and, by knowing and predicting which patients are destined to die and which to be saved, to keep himself free of blame.

2. Investigation in acute diseases[2] must be as follows. First, the patient's face, whether it resembles those of healthy people, and especially if it is like itself; this would be best, whereas if it was most opposite to its normal state, this would be most ominous: *viz.* the nose sharp, the eyes hollow, the temples sunken, the ears cold and contracted, with their lobes turned outward; and the skin about the forehead hard, taut, and parched, with the color of the whole face being yellow or even dark, and livid or leaden. Now if at the beginning of a disease the face is like this, and it is not yet possible to make judgments on the basis of other signs, you must inquire whether the patient has been sleepless, or whether there have been violent, moist movements of the cavity, or if he has been subject to a

[2] The acute diseases are pleurisy, pneumonia, ardent fever, and phrenitis. Cf. *Affections* 6, *Regimen in Acute Diseases* 5, *Airs Waters Places* 3, and *Prognostic* 4.

---

[4] καὶ πελιὸν ἢ μολιβδῶδες M: om. C′V
[5] οὖν MV: om. C′ [6] ἦν Kühlewein: εἴη C′: ᾖ MV

ἔχει αὐτόν. καὶ ἢν μέν τι τούτων ὁμολογῇ, ἧσσον νομίζειν δεινὸν εἶναι· κρίνεται δὲ τὰ τοιαῦτα[7] ἐν ἡμέρῃ τε καὶ νυκτί, ἢν διὰ ταύτας τὰς προφάσιας τὸ πρόσωπον τοιοῦτον ᾖ· ἢν δὲ μηδὲν τούτων φῇ εἶναι,[8] μηδ' ἐν τῷ χρόνῳ τῷ προειρημένῳ καταστῇ, εἰδέναι χρὴ ἐγγὺς ἐόντα τοῦ θανάτου.[9]

116 Ἢν δὲ καὶ παλαιοτέρου ἐόντος τοῦ νοσήματος ἢ τριταίου[10] | τὸ πρόσωπον τοιοῦτον ᾖ, περί τε τούτων ἐπανερέσθαι, περὶ ὧν καὶ πρότερον ἐκέλευσα καὶ τὰ ἄλλα σημεῖα σκέπτεσθαι, τά τε ἐν τῷ ξύμπαντι προσώπῳ[11] καὶ τὰ ἐν τοῖσιν ὀφθαλμοῖσιν. ἢν γὰρ τὴν αὐγὴν φεύγωσιν ἢ δακρύωσιν ἀπροαιρέτως ἢ διαστρέφωνται ἢ ὁ ἕτερος τοῦ ἑτέρου ἐλάσσων γίνηται ἢ τὰ λευκὰ ἐρυθρὰ ἴσχωσιν ἢ πελιὰ[12] ἢ φλέβια μέλανα ἐν ἑωυτοῖσιν ἔχωσιν, ἢ λῆμαι φαίνωνται περὶ τὰς ὄψιας ἢ καὶ[13] ἐναιωρεύμενοι ἢ ἐξίσχοντες ἢ ἔγκοιλοι ἰσχυρῶς γινόμενοι[14] ἢ τὸ χρῶμα τοῦ ξύμπαντος προσώπου ἠλλοιωμένον ᾖ, ταῦτα πάντα κακὰ νομίζειν εἶναι καὶ ὀλέθρια. σκοπεῖν δὲ χρὴ καὶ τὰς ὑποφάσιας τῶν ὀφθαλμῶν ἐν τοῖσιν ὕπνοισιν· ἢν γάρ τι ὑποφαίνηται τοῦ λευκοῦ ξυμβαλλομένων τῶν βλε-
118 φάρων, | μὴ ἐκ διαρροίης ἢ φαρμακοποσίης ἐόντι ἢ μὴ εἰθισμένῳ οὕτω καθεύδειν, φαῦλον τὸ σημεῖον καὶ θανατῶδες σφόδρα. ἢν δὲ καμπύλον γένηται ἢ πελιὸν ἢ ὠχρὸν[15] βλέφαρον ἢ χεῖλος ἢ ῥὶς μετά τινος τῶν

[7] τὰ τοιαῦτα M: ταῦτα C′: om. V
[8] εἶναι M: om. C′V

degree of hunger. If he answers in the affirmative to any of these questions, you may consider the case less dangerous: the crisis in such cases occurs after a day and a night, if the face is this way from any of these causes. But if the patient denies having any of these things, and no relief occurs in the time named, know that he is near death.

When the disease goes beyond three days with the face retaining these signs, make the same inquiries again which I asked for above, and pay attention to other signs of the whole face, and especially of the eyes. If they shun the light, or shed tears involuntarily, or turn awry, or one eye becomes smaller than the other one, or if their whites have red or livid marks in them or dark vessels, or rheums appear around the eyeballs, or if the eyes wander, protrude, or are very hollow, or the complexion of the whole face is altered, judge all these things to be bad and to presage death. You must also look for the eyes being partly visible during sleep; for if some part of the whites appears though the lids are mainly closed—unless this is due to diarrhea, or the administration of some purgative, or the person is habitually like this in his sleep—it is an indifferent sign and points strongly to death. If, in conjunction with one of the other signs, an eyelid becomes distorted or livid or

---

[9] *χρὴ . . . θανάτου* M: *τοῦτο τὸ σημεῖον θανατῶδες ἐόν* C′V [10] *ἢ τεταρταίου* add. M [11] *προσώπῳ* V: *σώματι* C′: *προσώπῳ καὶ τὰ ἐν τῷ σώματι* M

[12] *πελιὰ* MV: *πελιδνὰ* C′

[13] *ἢ καὶ* M: om. C′V

[14] Add. *ἢ αἱ ὄψιες αὐχμῶσαι καὶ ἀλαμπεῖς* M

[15] *ἢ ὠχρὸν* M: om. V: lacuna in C′

ἄλλων σημείων, εἰδέναι χρὴ ἐγγὺς ἐόντα τοῦ θανάτου· θανατῶδες δὲ καὶ χείλεα ὑπολυόμενα καὶ κρεμάμενα καὶ ψυχρὰ καὶ ἔκλευκα γινόμενα.

3. *Κεκλιμένον δὲ χρὴ καταλαμβάνεσθαι τὸν νοσέοντα ὑπὸ τοῦ ἰητροῦ ἐπὶ τὸ πλευρὸν τὸ δεξιὸν ἢ τὸ ἀριστερὸν καὶ τὰς χεῖρας καὶ τὸν τράχηλον καὶ τὰ σκέλεα ὀλίγον ἐπικεκαμμένα ἔχοντα καὶ τὸ ξύμπαν σῶμα ὑγρὸν κείμενον· οὕτω γὰρ καὶ οἱ πλεῖστοι τῶν ὑγιαινόντων κατακλίνονται· ἄρισται δέ εἰσι*[16] *τῶν κατακλίσεων αἱ ὅμοιαι*[17] *τῇσι τῶν ὑγιαινόντων. ὕπτιον δὲ κεῖσθαι καὶ τὰς χεῖρας καὶ τὸν τράχηλον*[18] *καὶ τὰ σκέλεα ἐκτεταμένα ἔχοντα ἧσσον ἀγαθόν. εἰ δὲ καὶ προπετὴς γένοιτο καὶ καταρρέοι ἀπὸ τῆς κλίνης ἐπὶ τοὺς πόδας, δεινότερόν ἐστιν.*[19] *εἰ δὲ καὶ γυμνοὺς τοὺς πόδας εὑρίσκοιτο ἔχων μὴ θερμοὺς κάρτα ἐόντας καὶ τὰς χεῖρας καὶ* | *τὸν τράχηλον*[20] *καὶ τὰ σκέλεα ἀνωμάλως διερριμμένα καὶ γυμνά, κακόν· ἀλυσμὸν γὰρ σημαίνει. θανατῶδες δὲ καὶ τὸ κεχηνότα καθεύδειν ἀεὶ καὶ τὰ σκέλεα ὑπτίου κειμένου ξυγκεκαμμένα εἶναι ἰσχυρῶς καὶ διαπεπλιγμένα.*[21] *ἐπὶ γαστέρα δὲ κεῖσθαι, ᾧ μὴ ξύνηθές ἐστι καὶ ὑγιαίνοντι οὕτω κοιμᾶσθαι,*[22] *παραφροσύνην τινὰ σημαίνει ἢ ὀδύνην τῶν περὶ τὴν κοιλίην*[23] *τόπων. ἀνακαθίζειν δὲ βούλεσθαι τὸν νοσέοντα τῆς νούσου ἀκμαζούσης πονηρὸν μὲν ἐν πᾶσι τοῖσιν ὀξέσι νοσήμασι, κάκιστον δ' ἐν τοῖσι περιπνευμονικοῖσιν. ὀδόντας δὲ πρίειν ἐν πυρετοῖς, ὁκόσοισι μὴ ξύνηθές ἐστιν ἀπὸ παίδων, μανικὸν καὶ*

120

yellow, or a lip or a nostril the same, you must be aware that death is near. It is also a fatal sign when the lips are slack, hanging, cold, and very white.

3. A patient ought to be found, when the physician arrives, reclining on his right or left side, holding his arms, neck and legs slightly bent, and his whole body in a relaxed position, for most healthy people lie this way, and the most favorable of postures are like those of the healthy. To lie on his back with his arms, neck and legs extended is less favorable. And if the patient lies bent forward, and slips down in his bed toward the foot, this position should provoke more unease. If he is found with naked feet that are not very warm, and his arms neck and legs cast about at random and naked, this is a bad sign, since it indicates distress; another fatal sign is if he sleeps with his mouth always wide open, and lies on his back with his legs tightly drawn up and twisted together. For a person to lie on his belly, if he does not habitually sleep in this position when healthy, indicates some mental disturbance, or pain in the parts around the cavity. For a patient to want to sit erect when his disease is at its crisis presages trouble in all acute diseases, but is most ominous in pneumonias. To grind the teeth in fevers, for those who have not had this habit since

---

16 *εἰσι* MV: om C′ 17 *ὅμοιαι* M: *ὁμοιόταται* C′V
18 *καὶ τὸν τράχηλὸν* M: om. C′V
19 Add. *τοῦτο ἐκείνου* C′
20 *καὶ τὸν τράχηλον* MV: om. C′
21 -*πλιγμένα* Littré (see vol. 3, p. xliv): -*πλεγμένα* codd.
22 *οὕτω κοιμᾶσθαι* MV: *κοιμᾶσθαι οὕτω, κακόν* C′
23 *κοιλίην* MV: *γαστέρα* C′

θανατῶδες· ἀλλὰ χρὴ προλέγειν ἐπ' ἀμφοῖν κίνδυνον
122 ἐσόμενον· ἢν δὲ καὶ παραφρονέων | τοῦτο ποιέῃ, ὀλέθριον γίνεται κάρτα ἤδη.

Ἕλκος δέ, ἤν τε καὶ προγεγονὸς τύχῃ ἔχων, ἤν τε καὶ ἐν τῇ νούσῳ γίνηται, καταμανθάνειν χρή. ἢν γὰρ μέλλῃ ἀπολεῖσθαι ὥνθρωπος,[24] πρὸ τοῦ θανάτου ἢ πελιδνόν τε καὶ ξηρὸν ἔσται ἢ ὠχρόν τε καὶ ξηρόν.[25]

4. Περὶ δὲ χειρῶν φορῆς τάδε γινώσκω· ὁκόσοισιν ἐν πυρετοῖσιν ὀξέσιν ἢ ἐν περιπλευμονίῃσι ἢ[26] ἐν φρενίτισι ἢ ἐν κεφαλαλγίῃσι πρὸ τοῦ προσώπου φερομένας καὶ θηρευούσας διὰ κενῆς, καὶ ἀποκαρφολογεούσας,[27] καὶ κροκύδας ἀπὸ τῶν ἱματίων ἀποτιλλούσας, καὶ ἀπὸ τοῦ τοίχου ἄχυρα ἀποσπώσας, πάσας εἶναι κακὰς καὶ θανατώδεας.

5. Πνεῦμα δὲ πυκνὸν μὲν ἐὸν πόνον σημαίνει ἢ φλεγμονὴν ἐν τοῖσιν ὑπὲρ τῶν φρενῶν χωρίοισι· μέγα δὲ ἀναπνεόμενον καὶ διὰ πολλοῦ χρόνου παραφροσύνην δηλοῖ·[28] ψυχρὸν δὲ ἐκπνεόμενον ἐκ τῶν ῥινῶν καὶ τοῦ στόματος ὀλέθριον κάρτα ἤδη γίνεται. εὔπνοιαν δὲ χρὴ νομίζειν κάρτα μεγάλην δύναμιν ἔχειν ἐς σωτηρίην ἐν πᾶσι τοῖσιν ὀξέσι νοσήμασιν, ὁκόσα ξὺν πυρετοῖσίν ἐστιν καὶ ἐν τεσσαράκοντα ἡμέρῃσι κρίνεται.

6. Οἱ δὲ ἱδρῶτες ἄριστοι μέν εἰσιν ἐν πᾶσι τοῖσιν
124 ὀξέσι νοσήμασιν, | ὁκόσοι ἂν ἐν ἡμέρῃσί τε κρισίμῃσι γένωνται καὶ τελείως τοῦ πυρετοῦ ἀπαλλάξωσιν. ἀγαθοὶ δὲ καὶ ὁκόσοι διὰ παντὸς τοῦ σώματος

childhood, points to raging and death, but the prediction of approaching danger must be made on the basis of both these signs (i.e., grinding of the teeth and mania). If the patients do this while they are delirious as well, the situation is already very dire.

Concerning a lesion, you must carefully investigate both whether the patient has had it before, and if it happened in a disease: for if the person is about to die, before death it will become livid and dry or greenish and dry.

4. Regarding movements with the hands, I am aware of the following. In patients with acute fevers, pneumonias, phrenitises, or headaches, for the hands to be raised in front of the face clutching at empty space, gathering up tiny particles, picking fluff off the bedding, or scratching bits of straw off the house wall are all bad, indeed fatal, signs.

5. Rapid breathing indicates trouble or inflammation in the parts above the diaphragm, while long, slow inspirations presage delirium. Cold expirations from the nostrils or mouth are already very dangerous. Healthy breathing must be considered to have great significance for survival in all acute diseases accompanied by fevers that reach a crisis by forty days.

6. The most favorable sweats in all acute diseases are those which occur on critical days and definitively end the fever. Good are also sweats that are present through the

---

[24] *ὤνθρωπος* MV: *ὁ ἀσθενῶν* C′ [25] *ὠχρόν τε καὶ ξηρόν* M: *ὀχρὸν καὶ σκληρόν* C′: *ξηρόν τε καὶ χλωρόν* V

[26] *ἢ* MV: *καὶ* C′ [27] *καὶ ἀποκαρφ.* transp. after *ἀποτιλλούσας* C′ [28] *δηλοῖ* MV: *σημαίνει* C′

γενόμενοι ἀπέδειξαν τὸν ἄνθρωπον εὐπετέστερον φέροντα τὸ νόσημα. οἳ δ' ἂν μὴ τούτων τι ἐξεργάσωνται,[29] οὐ λυσιτελέες. κάκιστοι δὲ οἱ ψυχροί τε καὶ μοῦνον περὶ τὴν κεφαλὴν[30] γινόμενοι καὶ τὸν αὐχένα· οὗτοι γὰρ σὺν μὲν ὀξεῖ πυρετῷ θάνατον προσημαίνουσιν, σὺν δὲ πρηϋτέρῳ, μῆκος νούσου. καὶ οἱ κατὰ πᾶν τὸ σῶμα ὡσαύτως γινόμενοι τοῖσι περὶ τὴν κεφαλήν· οἱ δὲ κεγχροειδέες καὶ μοῦνον περὶ τὸν τράχηλον γινόμενοι πονηροί· οἱ δὲ μετὰ σταλαγμῶν καὶ ἀτμίζοντες, ἀγαθοί. κατανοεῖν δὲ χρὴ τὸ σύνολον τῶν ἱδρώτων· γίνονται γὰρ οἱ μὲν δι' ἔκλυσιν, οἱ δὲ διὰ συντονίην φλεγμονῆς.

7. Ὑποχόνδριον δὲ ἄριστον μὲν ἀνώδυνόν τε ἐὸν

126 καὶ μαλθακὸν | καὶ ὁμαλὸν καὶ ἐπὶ δεξιὰ καὶ ἐπ' ἀριστερά· φλεγμαῖνον δὲ καὶ ὀδύνην παρέχον ἢ ἐντεταμένον ἢ ἀνωμάλως διακείμενον τὰ δεξιὰ πρὸς τὰ ἀριστερά, ταῦτα πάντα φυλάσσεσθαι χρή. εἰ δὲ καὶ σφυγμὸς ἐνείη ἐν τῷ ὑποχονδρίῳ, θόρυβον σημαίνει ἢ παραφροσύνην· ἀλλὰ τοὺς ὀφθαλμοὺς ἐπικατιδεῖν χρὴ τῶν τοιούτων· ἢν γὰρ αἱ ὄψιες πυκνὰ κινέωνται, μανῆναι τουτὸν ἐλπίς.

Οἴδημα δὲ ἐν τῷ ὑποχονδρίῳ σκληρόν τε ἐὸν καὶ ἐπώδυνον κάκιστον μέν, εἰ παρ' ἅπαν εἴη τὸ ὑποχόνδριον. εἰ δὲ εἴη ἐν τῷ ἑτέρῳ πλευρῷ, ἀκινδυνότερόν ἐστιν ἐν τῷ ἐπ' ἀριστερά. σημαίνει δὲ τὰ τοιαῦτα οἰδήματα ἐν ἀρχῇ μὲν κίνδυνον θανάτου ὀλιγοχρόνιον ἔσεσθαι· εἰ δὲ ὑπερβάλλοι εἴκοσιν ἡμέρας ὅ τε πυρετὸς ἔχων καὶ τὸ οἴδημα μὴ καθιστάμενον, ἐς διαπύη-

whole body, and that give evidence that the patient is bearing the disease quite well. Any sweats that do not have either of these effects are worthless. Worst are cold sweats that are present only in the head and neck, for these in conjunction with an acute fever indicate death, or with a mild fever that the course will be long. Similar are sweats present through the whole body in the same way as the ones in the head. Miliary sweats that are present only in the neck are difficult, but sweats accompanied by droplets that exhale vapor are good. Sweats must be carefully considered in all their aspects, for some arise with the resolution of an inflammation, and others through the intensification of an inflammation.

7. The hypochondrium is most favorable when it is painless, soft, and even on the right and left sides. But for it to be inflamed and painful, or contracted, or in different states on the right and the left, all these require careful attention. If there is also throbbing in the hypochondria, it presages confusion or delirium. Besides, in such cases the eyes must be observed, for if the eyeballs dart rapidly about, the patient may be expected to become delirious.

Swelling in the hypochondrium, when it is hard and painful, is worst when the whole hypochondrium is involved, but if only one side is swollen, this is less dangerous when on the left. Such swellings at the outset announce the danger of imminent death, but if the fever persists beyond twenty days and the swelling does not go down, it means a turn to internal suppuration. Such patients may

---

29 ἐξεργάσωνται MV: ἀπεργάζονται C′

30 Add. καὶ τὸ πρόσωπον MV

σιν τρέπεται. γίνεται δὲ τούτοισιν ἐν τῇ πρώτῃ περι-
128 όδῳ καὶ αἵματος ῥῆξις ἐκ τῶν ῥινῶν καὶ | κάρτα
ὠφελέει· ἀλλ' ἐπανερέσθαι χρή, εἰ τὴν κεφαλὴν ἀλγέουσιν ἢ ἀμβλυώσσουσιν· εἰ γάρ τι τοιοῦτον εἴη,
ἐνταῦθα ἂν ῥέποι. μᾶλλον δὲ τοῖσι νεωτέροισι πέντε
καὶ τριήκοντα ἐτέων τοῦ αἵματος ῥῆξιν προσδέχεσθαι
χρή.

Τὰ δὲ μαλθακὰ τῶν οἰδημάτων καὶ ἀνώδυνα καὶ
τῷ δακτύλῳ ὑπείκοντα χρονιωτέρας τὰς κρίσιας ποιέεται καὶ ἧσσον ἐκείνων δεινότερά ἐστιν· εἰ δὲ ὑπερβάλλοι ἑξήκοντα ἡμέρας ὅ τε πυρετὸς ἔχων καὶ τὸ
οἴδημα μὴ καθιστάμενον, ἔμπυον ἔσεσθαι σημαίνει
καὶ τοῦτο·[31] καὶ τὸ ἐν τῇ ἄλλῃ κοιλίῃ κατὰ τωὐτό·
ὁκόσα μὲν οὖν ἐπώδυνά τέ ἐστι καὶ σκληρὰ καὶ
μεγάλα, σημαίνει κίνδυνον θανάτου ὀλιγοχρόνιον,[32]
ὁκόσα δὲ μαλθακά τε καὶ ἀνώδυνα καὶ τῷ δακτύλῳ
πιεζεύμενα ὑπείκει, χρονιώτερα ἐκείνων.

Τὰς δὲ ἀποστάσιας ἧσσον τὰ ἐν τῇ γαστρὶ οἰδήματα ποιέεται τῶν ἐν τοῖσιν ὑποχονδρίοισιν, ἥκιστα δὲ τὰ ὑποκάτω τοῦ ὀμφαλοῦ ἐς ἀποπύησιν τρέ-
130 πεται· αἵματος δὲ | ῥῆξιν ἐκ τῶν ἄνω τόπων μάλιστα
προσδέχεσθαι. ἁπάντων δὲ χρὴ τῶν οἰδημάτων χρο-
νιζόντων περὶ ταῦτα τὰ χωρία ὑποσκέπτεσθαι τὰς
ἐμπυήσιας. τὰ δὲ διαπυήματα ὧδε χρὴ σκέπτεσθαι
τὰ ἐντεῦθεν· ὁκόσα μὲν ἔξω τρέπεται, ἄριστά ἐστιν ὡς
μάλιστα ἐκκλίνοντα καὶ ἐς ὀξὺ ἀποκυρτούμενα, τὰ δὲ
μεγάλα τε ἐόντα καὶ πλατέα καὶ ἥκιστα ἐς ὀξὺ ἀποκυρτούμενα, κάκιστα· ὁκόσα δὲ ἔσω ῥήγνυται, ἄρι-

also have hemorrhages from their nostrils in the first period, which brings considerable relief. But you must also ask them if they have headache or a dimness of their vision, since if either of these is present, the disease will migrate in that direction (i.e., toward the head). Such an epistaxis is more likely to occur in people younger than thirty-five years.

Soft swellings without pain, that yield to the (sc. examining) finger, have their crises later, and are less serious than the ones just described. When the fever has persisted longer than sixty days without the swelling going down, this too indicates there will be internal suppuration. In the rest of the cavity it is the same, too, for swellings that are painful, hard, and widespread announce the danger of imminent death, while those that are soft, painless, and which remain indented when they are pressed by a finger are more chronic than the others.

Swellings in the belly produce abscessions less often than those in the hypochondrium, and those below the navel turn least often into suppurations: in these expect a hemorrhage, most likely from the upper parts. You must keep a look out for suppurations in all swellings that last a long time in these parts, and observe the suppurations as follows. Those which turn outward are most favorable if they have convex sides and come to a point; ones that are large, flat, and fail to come to a point are the worst. Sup-

---

[31] This punctuation is according to Jouanna; previously the sentence division was made before *καὶ τοῦτο*.

[32] *θανάτου ὀλιγοχρόνιον* Aldina: *θανάτου ὀλιγοχρονίου* C′M: *καὶ θανάτους ὀλιγοχρονίους* V

στά ἐστιν, ἃ τῷ ἔξω χωρίῳ μηδὲν ἐπικοινωνέει, ἀλλ' ἔστιν προσεσταλμένα τε καὶ ἀνώδυνα καὶ πᾶν τὸ ἔξω χωρίον ὁμόχροον φαίνηται. τὸ δὲ πύον ἄριστον λευκόν τε εἶναι καὶ ὁμαλὸν[33] καὶ λεῖον καὶ ὡς ἥκιστα δυσῶδες· τὸ δὲ ἐναντίον τούτου κάκιστον.

8. Οἱ δὲ ὕδρωπες οἱ ἐκ τῶν ὀξέων νοσημάτων πάντες κακοί· οὔτε γὰρ τοῦ πυρὸς ἀπαλλάσσουσιν ἐπώδυνοί τέ εἰσιν καὶ θανατώδεες. ἄρχονται δὲ οἱ πλεῖστοι ἀπὸ τῶν κενεώνων καὶ τῆς ὀσφύος, οἱ δὲ καὶ ἀπὸ τοῦ ἥπατος. οἷσι μὲν οὖν ἀπὸ τῶν κενεώνων αἱ ἀρχαὶ γίνοντα τῶν ὑδρώπων, οἵ τε πόδες οἰδέουσι, καὶ διάρροιαι πολυχρόνιοι ἔχουσιν, οὔτε τὰς ὀδύνας λύουσαι τὰς ἐκ τῶν κενεώνων τε καὶ τῆς ὀσφύος οὔτε τὴν γαστέρα λαπάσσουσαι. | ὁκόσοισι δὲ ἀπὸ τοῦ ἥπατος ὕδρωπες[34] γίνονται, βῆξαί τε θυμὸς ἐγγίνεται αὐτοῖσι καὶ ἀποπτύουσιν οὐδὲν ἄξιον λόγου καὶ οἱ πόδες οἰδέουσιν καὶ ἡ γαστὴρ οὐ διαχωρέει, εἰ μὴ σκληρά τε καὶ πρὸς ἀνάγκην, καὶ περὶ τὴν κοιλίην γίνεται οἰδήματα, τὰ μὲν ἐπὶ δεξιά, τὰ δὲ ἐπ' ἀριστερά, ἱστάμενά τε καὶ καταπαυόμενα.

132

9. Κεφαλὴ δὲ καὶ χεῖρες καὶ πόδες ψυχρὰ ἐόντα κακὸν τῆς τε κοιλίης καὶ τῶν πλευρῶν θερμῶν ἐόντων. ἄριστον δὲ ἅπαν τὸ σῶμα θερμόν τε εἶναι καὶ μαλθακὸν ὁμαλῶς.

Στρέφεσθαι δὲ χρὴ ῥηϊδίως τὸν νοσέοντα καὶ ἐν τοῖσι μετεωρισμοῖσιν ἐλαφρὸν εἶναι· εἰ δὲ βαρὺς ἐὼν φαίνοιτο καὶ τὸ ἄλλο σῶμα καὶ τὰς χεῖρας καὶ τοὺς πόδας, ἐπικινδυνότερόν ἐστιν. εἰ δὲ πρὸς τῷ βάρει καὶ

purations that rupture inward are best if they have no communication with the outside, but are compact, painless, and of uniform color on the surface. The most favorable pus is white, uniform, smooth and the least foul-smelling: pus opposite to this is worst.

8. Dropsies arising out of acute diseases are all bad, since they do not allay the fever, and are themselves painful and deadly. Most begin from the flanks and the lower back, although others start from the liver. In dropsies beginning from the flanks, the feet swell, and chronic diarrhea supervenes which neither relieves the pains in the flanks and loins nor softens the belly. In cases where dropsy begins from the liver, the patient feels a need to cough, but fails to expectorate anything worth mentioning, his feet swell, and from his belly nothing is passed but what is hard and forced (i.e., by a purgative), and around the cavity swellings develop, some on the right and others on the left, which alternately increase and recede.

9. For the head, hands and feet to be cold while the cavity and sides are warm is a bad sign: best is for the whole body to be uniformly warm and soft.

A patient must be able to turn himself easily, and seem light when he is lifted. If, however, his body in general seems heavy, but especially the hands and feet, it is a more dangerous sign. If in addition to the heaviness, his nails

---

[33] *καὶ ὁμαλὸν* C′V: om. M

[34] *ὕδρωπες* MV: om. C′

οἱ ὄνυχες καὶ οἱ δάκτυλοι πελιδνοὶ γίνονται, προσδόκιμος ὁ θάνατος αὐτίκα· μελαινόμενοι δὲ παντελῶς οἱ δάκτυλοι ἢ οἱ πόδες ἧσσον ὀλέθριοι τῶν πελιδνῶν· ἀλλὰ τἆλλα σημεῖα σκέπτεσθαι χρή· ἢν γὰρ εὐπετέως φέρων φαίνηται τὸ κακὸν καὶ ἄλλο τι τῶν περι-
134 εστικῶν πρὸς τουτοῖσι τοῖσι σημείοισι | ὑποδεικνύῃ, τὸ νόσημα ἐς ἀπόστασιν τραπῆναι ἐλπίς, ὥστε τὸν μὲν ἄνθρωπον περιγενέσθαι, τὰ δὲ μελανθέντα τοῦ σώματος ἀποπεσεῖν.

Ὄρχιες δὲ καὶ αἰδοῖον ἀνεσπασμένα πόνους ἰσχυροὺς σημαίνει καὶ κίνδυνον θανατώδεα.

10. Περὶ δε ὕπνων ὥσπερ κατὰ φύσιν ξύνηθές ἡμῖν ἐστι, τὴν μὲν ἡμέρην ἐγρηγορέναι χρή, τὴν δὲ νύκτα καθεύδειν· εἰ δὲ εἴη τοῦτο μεταβεβλημένον, κάκιον γίνεται· ἥκιστα δ᾽ ἂν λυπέοιτο, εἰ κοιμῷτο τὸ πρωῒ ἐς τὸ τρίτον μέρος τῆς ἡμέρης· οἱ δ᾽ ἀπὸ τούτου τοῦ χρόνου ὕπνοι πονηρότεροί εἰσι. κάκιστον δὲ μὴ κοιμᾶσθαι μήτε τῆς νυκτὸς μήτε τῆς ἡμέρης· ἢ γὰρ ὑπὸ ὀδύνης τε καὶ πόνων ἀγρυπνοίη ἂν ἢ παραφρονήσει[35] ἀπὸ τούτου τοῦ σημείου.

11. Διαχώρημα δὲ ἄριστόν ἐστιν μαλθακόν τε καὶ ξυνεστηκός, καὶ τὴν ὥρην ἥνπερ καὶ ὑγιαίνοντι διεχώρει, πλῆθος δὲ πρὸς λόγον τῶν ἐσιόντων· τοιαύτης γὰρ ἐούσης τῆς διεξόδου ἡ κάτω κοιλίη ὑγιαίνοι ἄν. εἰ δὲ εἴη ὑγρὸν τὸ διαχώρημα, ξυμφέρει μήτε τρύζειν
136 μήτε πυκνόν τε καὶ κατ᾽ ὀλίγον | διαχωρέειν· κοπιῶν γὰρ ὁ ἄνθρωπος ὑπὸ τῆς ξυνεχέος ἐξαναστάσιος καὶ ἀγρυπνοίη ἄν· εἰ δὲ ἀθρόον πολλάκις διαχωρέοι, κίν-

and fingers become livid, death is to be expected soon. But for the fingers and toes to become completely black is less dangerous than for them to become livid. Attention must be paid to other signs as well: if a person seems to be bearing his disease well, or he displays some other sign indicating survival in addition to the ones mentioned before, his disease may be expected to turn toward an abscession, with the result that he survives; but the blackened parts will be lost.

For the testicles and penis to be retracted foreshadows severe pains and mortal danger.

10. As for sleep, it should retain our customary natural pattern: to be awake during the day and to sleep at night. If this order is interrupted, it bodes ill. Least damage would result if a person slept during the early forenoon through the first third of the day: to sleep after that indicates trouble. Worst is to sleep during neither the day nor the night, for either such a person is sleepless because of pain and distress, or delirium will develop after this sign.

11. The most favorable stools are soft but solid, pass at the same time they do in a healthy person, and are in their quantity proportionate to what has been ingested. If the movement is like this, it indicates that the lower cavity is healthy. If, however, the excreta are moist, it is important that they make no sound, and that they do not pass frequently a little at a time. For if someone is wearied by having to get up frequently from bed, he will also lack sleep; and, if the stools pass frequently in a great mass,

---

[35] παραφρονήσει V: παραφροσύνη ἔσται C′M

δυνος λειποθυμῆσαι. ἀλλὰ χρὴ κατὰ τὸ πλῆθος τῶν ἐσιόντων ὑποχωρέειν δὶς ἢ τρὶς τῆς ἡμέρης καὶ τῆς νυκτὸς ἅπαξ· τὸ δὲ πλεῖστον ὑπίτω τὸ πρωΐ, ὥσπερ καὶ ξύνηθες ἦν τῷ ἀνθρώπῳ. παχύνεσθαι δὲ χρὴ τὸ διαχώρημα πρὸς τὴν κρίσιν ἰούσης τῆς νούσου. ὑπόπυρρον δὲ ἔστω καὶ μὴ λίην δυσῶδες· ἐπιτήδειον δὲ καὶ ἕλμινθας στρογγύλας διεξιέναι μετὰ τοῦ διαχωρήματος πρὸς τὴν κρίσιν ἰούσης τῆς νούσου. δεῖ δὲ ἐν παντὶ νοσήματι λαπαρὴν τὴν κοιλίην εἶναι καὶ εὔογκον. ὑδαρὲς δὲ κάρτα ἢ λευκὸν ἢ χλωρὸν[36] ἰσχυρῶς ἢ ἀφρῶδες διαχωρέειν, πονηρὰ ταῦτα πάντα. πο-
138 νηρὸν δὲ καὶ σμικρόν τε ἐὸν καὶ γλίσχρον καὶ | λεῖον καὶ λευκὸν καὶ ὑπόχλωρον. τούτων δὲ[37] θανατωδέστερα ἂν εἴη τὰ μέλανα ἢ λιπαρὰ ἢ πελιὰ ἢ ἰώδεα[38] καὶ κάκοσμα.[39] τὰ δὲ ποικίλα χρονιώτερα μὲν τούτων. ὀλέθρια δὲ οὐδὲν ἧσσον· ἔστι δὲ ταῦτα ξυσματώδεα καὶ χολώδεα καὶ πρασοειδέα καὶ μέλανα, ποτὲ μὲν ὁμοῦ διεξερχόμενα ἀλλήλοισι, ποτὲ δὲ κατὰ μέρος.

Φῦσαν δὲ ἄνευ ψόφου καὶ πραδήσιος διεξιέναι ἄριστον· κρέσσον δὲ καὶ σὺν ψόφῳ διεξελθεῖν ἢ αὐτοῦ ἀνειλέεσθαι· καίτοι καὶ οὕτω διελθοῦσα σημαίνει πονέειν τι τὸν ἄνθρωπον ἢ παραφρονέειν, ἢν μὴ

[36] Add. ἢ ἐρυθρόν MV

[37] δὲ C′M: δ' ἔτι V

[38] τὰ μέλανα . . . ἰώδεα M: μέλανα ἢ πελιδνὰ ἢ λιπαρὰ ἢ ἰώδη C′: πελιὰ ἢ μέλανα ἢ λιπαρὰ ἢ ἰώδεα V

[39] Jouanna adds a passage to the text here on the basis of *Coan Prenotions* 621: καὶ τὰ χολώδεα ἔχοντα ἐν ἑωυτοῖσι φακῶν ἢ

there is a danger he will faint. Ideally, stools should be passed in proportion to what is eaten, two or three times during the day and once at night: the greatest quantity should descend early in the morning, as is habitual for a person (sc. in health). The stools should become thicker as a disease moves toward its crisis; they should be somewhat flame-colored, and not excessively malodorous. It is also advantageous if round worms are expelled with the motions as a disease approaches its crisis. In every disease the cavity should be soft and of the proper size. To pass overly liquid, white, very yellow, or foamy stools, are all signs of trouble. Also troublesome are scanty, viscous, smooth, white and yellowish excreta. But more deadly than these would be black, greasy, livid, or verdigris-colored, malodorous stools.[3] Variegated stools indicate a more chronic disease than these, without being any less ominous; such contain shreds of flesh, and are bilious, leek-colored, and dark—sometimes appearing with all these features together simultaneously, at other times in succession.

To pass wind without any noise or breaking sound is best. But it is still better for it to come out accompanied by a noise than for it to be pent up inside. Nevertheless, if it does pass like this, it indicates that the person is suf-

[3] Jouanna adds "and bilious stools which contain material like pounded lentils or chick-peas, or like fresh clots of blood, and in their smell resemble the stools of infants" based on the text of *Coan Prenotions* 621.

ἐρεβίνθων ἐρίγμασι παραπλήσια, ἢ οἷον θρόμβους αἵματος εὐανθεῖς, κατὰ τὴν ὀδμὴν ὅμοια τοῖσι τῶν νηπίων.

ἑκὼν οὕτω ποιέηται ὁ ἄνθρωπος τὴν ἄφεσιν τῆς φύσης. τοὺς δὲ ἐκ τῶν ὑποχονδρίων πόνους τε καὶ τὰ κυρτώματα, ἢν ᾖ νεαρὰ καὶ μὴ ξὺν φλεγμονῇ, λύει βορβορυγμὸς ἐγγενόμενος ἐν τῷ ὑποχονδρίῳ καὶ μάλιστα μὲν διεξελθὼν ξὺν κόπρῳ τε καὶ οὔρῳ· εἰ δὲ μή, καὶ αὐτὸς διαπεραιωθείς· ὠφελέει δὲ καὶ ὑποκαταβὰς ἐς τὰ κάτω χωρία.

12. *Οὖρον δὲ ἄριστόν ἐστιν, ὅταν ᾖ λευκή τε ἡ ὑπόστασις καὶ λείη καὶ ὁμαλὴ παρὰ πάντα τὸν χρόνον, ἔστ' ἂν κριθῇ ἡ* | *νοῦσος· σημαίνει γὰρ ἀσφάλειάν τε καὶ νόσημα ὀλιγοχρόνιον ἔσεσθαι. εἰ δὲ διαλείποι καὶ ποτὲ μὲν καθαρὸν οὐρέοιτο,*[40] *ποτὲ δὲ ὑφίσταται τὸ λευκόν τε καὶ λεῖον, χρονιωτέρη γίνεται ἡ νοῦσος καὶ ἧσσον ἀσφαλής. εἰ δὲ εἴη τό τε οὖρον ὑπέρυθρον καὶ ἡ ὑπόστασις ὑπέρυθρός τε καὶ λείη, πολυχρονιώτερον μὲν τοῦτο τοῦ πρώτου*[41] *γίνεται, σωτήριον δὲ κάρτα. κριμνώδεες δὲ ἐν τοῖσιν οὔροισιν ὑποστάσιες πονηραί· τούτων δ' ἔτι κακίους αἱ πεταλώδεες· λεπταὶ δὲ καὶ λευκαὶ κάρτα φλαῦραι· τούτων δ' ἔτι κακίους αἱ πιτυρώδεες. νεφέλαι δὲ ἐμφερόμεναι*[42] *τοῖσιν οὔροισι λευκαὶ μὲν ἀγαθαί, μέλαιναι δὲ φλαῦραι. ἔστ' ἂν δὲ πυρρὸν ᾖ τὸ οὖρον καὶ λεπτόν,*[43] *ἄπεπτον σημαίνει τὸ νόσημα εἶναι· εἰ δὲ καὶ πολυχρόνιον εἴη,*[44] *τοιοῦτον ἐόν, κίνδυνος μὴ οὐ δυνήσεται ὁ ἄνθρωπος διαρκέσαι, ἔστ' ἂν πεπανθῇ ἡ νοῦσος.* | *θανατωδέ-*

140

142

[40] *οὐρέοιτο* V: *οὐρέοι* C'M
[41] *πρώτου* MV: *προτέρου* C'

fering from some disorder or is delirious, unless he is expelling the wind deliberately. Pains originating from the hypochondria that are accompanied by tympanites: if they are of recent origin and without inflammation, they will be relieved by rumbling in the hypochondrium, especially when the wind is expelled along with stools and urine, but if not, then even when it is expelled alone. It also brings some benefit if the wind descends into the lower regions.

12. Urine is most favorable when its sediment is white, fine, and of a uniform consistency for the whole time until the disease has its crisis, for this indicates safety, and that the disease will be of short duration. If urine passes irregularly, and at one time is clear but at another time precipitates a white, fine sediment, such a disease will be longer and less secure. If the urine is reddish with a fine, reddish sediment, such a disease will be longer than the first one, but is still quite safe. Sediments in urines like coarse meal are a bad sign; and even worse than these are urines filled with flakes. Thin, white sediments (sc. of this kind) are a very indifferent sign, and even worse than these are sediments like bran. Clouds suspended in urine, if white, are a good sign, but if dark, they are indifferent. As long as urine is flame-colored and thin, it indicates that a disease is unconcocted. If such a disease lasts for a long time (sc. with urine) like this, there is a danger that the patient will not be able to hold out until the disease arrives

---

[42] *ἐμφερόμεναι* M: *ἐνεωρούμεναι* C′: *ἐναιωρεύμεναι* V

[43] *πυρρόν . . . λεπτόν* M: *λεπτὸν . . . πυρρὸν καὶ ὁμαλόν* C′: *ὑποπυρρόν . . . λεπτόν* V

[44] Add. *τὸ νόσημα, τὸ δὲ οὖρον* C′

στερα δὲ τῶν οὔρων τά τε ὑδατώδεα καὶ δυσώδεα καὶ μέλανα καὶ παχέα· ἔστι δὲ τοῖσι μὲν ἀνδράσι καὶ τῇσι γυναιξὶ τὰ μέλανα τῶν οὔρων κάκιστα, τοῖσι δὲ παιδίοισι τὰ ὑδατώδεα. ὁκόσοι δ' ἂν οὖρα λεπτὰ καὶ ὠμὰ οὐρέωσι πολὺν χρόνον, ἢν τἆλλα σημεῖα ὡς περιεσομένοισι ᾖ, τούτοισιν ἀπόστασιν δεῖ προσδέχεσθαι ἐς τὰ κάτω τῶν φρενῶν χωρία. καὶ τὰς λιπαρότητας δὲ τὰς ἄνω ἐφισταμένας ἀραχνοειδέας μέμφεσθαι· ξυντήξιος γὰρ σημεῖα. σκοπεῖν δὲ χρὴ τῶν οὔρων, ἐν οἷσιν αἱ νεφέλαι, ἤν τε κάτω ἔωσιν ἤν τε ἄνω, καὶ τὰ χρώματα ὁκοῖα ἴσχουσιν· καὶ τὰς μὲν κάτω φερομένας σὺν τοῖσι χρώμασιν, οἷσιν εἴρηται ἀγαθὰς εἶναι, καὶ ἐπαινέειν, τὰς δὲ ἄνω σὺν τοῖσι χρώμασιν, οἷσιν εἴρηται κακὰς εἶναι, καὶ μέμφεσθαι. μὴ ἐξαπατάτω δέ σε, ἤν τι αὐτὴ ἡ κύστις νόσημα ἔχουσα τῶν οὔρων τοιαῦτα ἀποδιδῷ· οὐ γὰρ τοῦ ὅλου σώματος σημεῖόν ἐστιν, ἀλλ' αὐτῆς καθ' ἑωυτήν.

13. Ἔμετος δὲ ὠφελιμώτατος φλέγματός τε καὶ χολῆς συμμεμιγμένων | ὡς μάλιστα· καὶ μὴ παχὺς μηδὲ πολὺς κάρτα ἐμείσθω· οἱ δὲ ἀκρητέστεροι κακίους. εἰ δὲ εἴη τὸ ἐμεύμενον πρασοειδὲς ἢ πελιὸν ἢ μέλαν, ὅ τι ἂν ᾖ τούτων τῶν χρωμάτων, νομίζειν χρὴ πονηρὸν εἶναι· εἰ δὲ καὶ πάντα τὰ χρώματα ὁ αὐτὸς ἄνθρωπος ἐμέοι, κάρτα ὀλέθριον ἤδη γίνεται· τάχιστον δὲ θάνατον σημαίνει τὸ πελιὸν τῶν ἐμεσμάτων, εἰ ὄζοι δυσῶδες· πᾶσαι δὲ αἱ ὑπόσαπροι καὶ δυσώδεες ὀδμαὶ κακαὶ ἐπὶ πᾶσι τοῖσιν ἐμεομένοισι.

14. Πτύελον δὲ χρὴ ἐπὶ πᾶσι τοῖσιν ἀλγήμασι

144

at concoction. Particularly fatal of urines are the watery, foul-smelling, dark, and thick—for men and women dark urines are the worst, for children the watery ones. Persons who pass thin, unconcocted urines for a long time, if their other signs suggest recovery, you should expect to have an apostasis in the regions below the diaphragm. Also condemn fatty material like cobwebs floating on the surface (sc. of urine), since this is a sign of wasting. Examine urines containing clouds to see whether these are near the bottom or the top, and what colors they have: clouds suspended near the bottom, if they have the colors mentioned above as favorable, are to be applauded, whereas those near the top, having the colors noted as unfavorable, should be condemned. Do not let yourself be deceived in a case where the bladder itself produces such urines due to some disease of its own, for this does not reflect on the body as a whole, but only on the bladder itself.

13. Vomitus is most favorable if is very thoroughly mixed with phlegm and bile; it should not be brought up either too thick or in too great a quantity: less mixed it is a worse indication. If what is vomited up is leek-colored, livid, or dark, any of these colors must be taken as bad; furthermore, if the same person vomits material with all these colors, his state is already very serious. The earliest death is indicated by livid vomitus if it smells fetid; in fact, odors tending toward the putrid or the foul are bad in every vomitus.

14. Sputum expectorated from any painful condition in

τοῖσι περὶ τὸν πλεύμονά τε καὶ τὰς πλευρὰς ταχέως τε ἀναπτύεσθαι καὶ εὐπετέως, ξυμμεμιγμένον τε φαίνεσθαι τὸ ξανθὸν ἰσχυρῶς τῷ πτυέλῳ· εἰ γὰρ πολλῷ ὕστερον μετὰ τὴν ἀρχὴν τῆς ὀδύνης ἀναπτύοιτο ξανθὸν ἐὸν ἢ πυρρὸν ἢ πολλὴν βῆχα παρέχον ἢ μὴ ἰσχυρῶς ξυμμεμιγμένον, κάκιον γίνεται· τό τε γὰρ
146 ξανθὸν | ἄκρητον ἐὸν κινδυνῶδες, τὸ δὲ λευκὸν καὶ γλίσχρον καὶ στρογγύλον ἀλυσιτελές· κακὸν δὲ καὶ χλωρόν τε ἐὸν κάρτα καὶ ἀφρῶδες· εἰ δὲ εἴη οὕτως ἄκρητον, ὥστε καὶ μέλαν φαίνεσθαι, δεινότερόν ἐστι τοῦτο ἐκείνων· κακὸν δὲ κὴν μηδὲν ἀνακαθαίρηται μηδὲ προΐῃ ὁ πλεύμων, ἀλλὰ πλήρης ἐὼν ζέῃ ἐν τῇ φάρυγγι. κορύζας δὲ καὶ πταρμοὺς ἐπὶ πᾶσι τοῖσι περὶ τὸν πλεύμονα νοσήμασι καὶ προγεγονέναι καὶ ἐπιγενέσθαι, κακόν, ἀλλ᾽ ἐν τοῖσιν ἄλλοισι νοσήμασι τοῖσι θανατώδεσιν οἱ πταρμοὶ λυσιτελέες. αἵματι δὲ ξυμμεμιγμένον μὴ πολλῷ πτύελον ξανθὸν ἐν τοῖσι περιπλευμονικοῖσιν ἐν ἀρχῇ μὲν τῆς νούσου πτυόμενον περιεστικὸν κάρτα· ἑβδομαίῳ δὲ ἐόντι ἢ παλαιοτέρῳ ἧσσον ἀσφαλές. πάντα δὲ τὰ πτύελα πονηρά ἐστιν, ὁκόσα ἂν τὴν ὀδύνην μὴ παύῃ· κάκιστα δὲ τὰ μέλανα, ὡς διαγέγραπται· τὰ παύοντα δὲ τὴν ὀδύνην πάντα ἀμείνω πτυόμενα.

15. Ὁκόσα δὲ τῶν ἀλγημάτων ἐκ τούτων τῶν χω-
148 ρίων μὴ | παύηται μήτε πρὸς τὰς τῶν πτυέλων καθάρσιας μήτε πρὸς τὴν τῆς κοιλίης ἐκκόπρωσιν μήτε πρὸς τὰς φλεβοτομίας τε καὶ διαίτας, εἰδέναι δεῖ ἐκπυήσοντα. τῶν δὲ ἐμπυημάτων[45] ὁκόσα μὲν ἔτι χολώ-

the region of the lungs and sides must be coughed up quickly and easily; and its yellow component must be thoroughly mixed with the rest. For if only long after the onset of pain yellow or flame-colored sputum is expectorated, or it provokes violent coughing, or it is not well-mixed, this is quite a bad sign; unmixed yellow sputum is dangerous, and a white, viscous, lumpy one brings no advantage. Also bad is very green, foamy (sc. sputum); and if it is so unmixed that it has a dark appearance, this is more threatening than the other sputa mentioned. It is also bad if nothing is cleaned upward, and the lung fails to expel anything, but it is full and sends foam into the throat. Nasal mucus and sneezing are bad signs in all diseases located around the lung, whether they were already present before or they develop later; in all other dangerous diseases, though, sneezing is beneficial. In pneumonias yellow sputum mixed with not too much blood when brought up at the beginning of the disease is a very good sign of recovery; on the seventh day or later it is less reassuring. All sputa that do not relieve the pain are bad, but worst are the dark-colored, as has been explained, whereas all sputa that do relieve the pain are better signs.

15. If pains from these parts (i.e., the lungs and sides) are not relieved, although sputum is expectorated, stools are passed from the cavity, blood is let, and dietetic measures are applied, this indicates impending suppuration. Suppurations that occur while the sputum is still bilious

---

45 ἐμπ. C′V: ἐκπ. M

δεος ἐόντος τοῦ πτυέλου ἐκπυΐσκεται, ὀλέθρια κάρτα, εἴτε ἐν μέρει τὸ χολῶδες τῷ πύῳ ἀναπτύοιτο εἴτε ὁμοῦ. μάλιστα δέ, ἢν ἄρξηται χωρέειν τὸ ἐμπύημα[46] ἀπὸ τούτου τοῦ πτυέλου, ἑβδομαίου ἐόντος τοῦ νοσήματος· ἐλπὶς δὲ τὸν τὰ τοιαῦτα πτύοντα ἀποθανεῖσθαι τεσσαρεσκαιδεκαταῖον, ἢν μή τι αὐτῷ ἐπιγένηται <κακὸν ἢ>[47] ἀγαθόν. ἔστι δὲ τὰ μὲν ἀγαθὰ τάδε· εὐπετέως φέρειν τὸ νόσημα, εὔπνοον εἶναι, τῆς ὀδύνης ἀπηλλάχθαι, τὸ πτύελον ῥηϊδίως ἀναβήσσειν, τὸ σῶμα πᾶν ὁμαλῶς θερμόν τε εἶναι καὶ μαλθακὸν καὶ δίψαν μὴ ἔχειν, οὖρά τε καὶ διαχωρήματα καὶ ὕπνους καὶ ἱδρῶτας, ὡς διαγέγραπται ἕκαστα εἰδέναι ἀγαθὰ ἐόντα ἐπιγενέσθαι· οὕτω μὲν γὰρ πάντων τούτων[48] ἐπιγενομένων οὐκ ἂν ἀποθάνοι ὁ ἄνθρωπος· ἢν δὲ τὰ μὲν τούτων ἐπιγένηται, τὰ δὲ μή,[49] πλείονα χρόνον |
150 ζήσας ἢ τεσσαρεσκαίδεκα ἡμέρας ἀπόλοιτ' ἄν. κακὰ δὲ τἀναντία τούτων· δυσπετέως φέρειν τὴν νοῦσον, πνεῦμα μέγα καὶ πυκνὸν εἶναι, τὴν ὀδύνην μὴ πεπαῦσθαι, τὸ πτύελον μόλις ἀναβήσσειν, διψῆν κάρτα, τὸ σῶμα ὑπὸ τοῦ πυρὸς ἀνωμάλως ἔχεσθαι καὶ τὴν μὲν κοιλίην καὶ τὰς πλευρὰς θερμὰς εἶναι ἰσχυρῶς, τὸ δὲ μέτωπον καὶ τὰς χεῖρας καὶ τοὺς πόδας ψυχρούς, οὖρα δὲ καὶ διαχωρήματα καὶ ὕπνους καὶ ἱδρῶτας, ὡς διαγέγραπται ἕκαστα εἰδέναι κακὰ ἐόντα <ἐπιγίνεσθαι>.[50] εἰ δε οὕτως ἐπιγίνοιτό τι τῷ πτυέλῳ τούτῳ, ἀπόλοιτ' ἂν ὁ ἄνθρωπος, πρὶν ἢ ἐς τὰς τεσσαρεσκαί-

46 ἐμπ. MV: ἐκπ. C'

are very dangerous, whether the bilious expectoration occurs alternately or together with the formation of pus—especially when the pus begins to form out of this (sc. bilious) sputum on the seventh day of the disease. It is to be expected that anyone producing such sputum will perish on the fourteenth day, unless some ‹bad or› good sign reveals itself in him. The good signs are: bearing the condition without strain; breathing freely; becoming free of pain; coughing up sputum easily; the whole body being evenly warm and relaxed; the absence of thirst; the urines, stools, sleep, and sweats all being such as was indicated to be known to be favorable. For if all these indications are such, the person will not die: however, if some but not others of the signs that appear are propitious, the person will live longer than fourteen days, before he dies. The bad signs are the opposite of these: to be strained by the disease; to breathe deeply and rapidly; to suffer persistent pain; to cough up sputum only with difficulty; to suffer severe thirst; for the body to be unevenly affected with fever, and the belly and sides to be very warm, while the forehead, hands, and feet are cold; the urines, stools, sleep, and sweats to be such as I indicated above to be know to be unfavorable. If any of these appears in conjunction with such a (sc. bilious) sputum, the patient will succumb before arriving at the fourteenth day, on either

---

[47] Add. Jouanna; cf. *Coan Prenotions* 386
[48] Add. *τῶν σημείων* C′
[49] *τὰ μὲν τούτων . . . τὰ δὲ μή* C′: *τὰ μέν τι αὐτῶν . . . τὰ δὲ μή* M: *τὸ μέν τι αὐτῶν . . . τὸ δὲ μή* V
[50] Add. Littré

δεκα ἡμέρας ἀφικέσθαι, ἢ ἐναταῖος ἢ ἑνδεκαταῖος. οὕτως οὖν ξυμβάλλεσθαι χρή, ὡς τοῦ πτυέλου τούτου θανατώδεος ἐόντος μάλα καὶ οὐ περιάγοντος ἐς τὰς τεσσαρεσκαίδεκα ἡμέρας. τὰ δὲ ἐπιγινόμενα ἀγαθά τε καὶ κακὰ ξυλλογιζόμενον ἐκ τούτων χρὴ τὰς προρρήσιας ποιέεσθαι· οὕτω γὰρ ἄν τις ἀληθεύοι μάλιστα. αἱ δὲ ἄλλαι ἐκπυήσιες αἱ πλεῖσται ῥήγνυνται, αἱ μὲν εἰκοσταῖαι, αἱ δὲ τριηκοσταῖαι, αἱ δὲ τεσσαρακονθήμεροι, αἱ δὲ πρὸς τὰς ἑξήκοντα ἡμέρας ἀφικνέονται.

16. Ἐπισκέπτεσθαι δὲ χρὴ τὴν ἀρχὴν τοῦ ἐμπυ-
152 ήματος ἔσεσθαι | λογιζόμενον ἀπὸ τῆς ἡμέρης, ᾗ τὸ πρῶτον ὁ ἄνθρωπος ἐπύρεξεν ἢ εἴ ποτε αὐτὸν ῥῖγος ἔλαβεν καὶ εἰ φαίη ἀντὶ τῆς ὀδύνης αὐτῷ βάρος ἐγγενῆσθαι ἐν τῷ τόπῳ, ᾧ ἤλγει· ταῦτα γὰρ ἐν ἀρχῇσι γίνεται τῶν ἐμπυημάτων. ἐξ οὖν τούτων τῶν χρόνων τὴν ῥῆξιν χρὴ προσδέχεσθαι τοῦ πύου ἔσεσθαι ἐς τοὺς χρόνους τοὺς προειρημένους. εἰ δὲ εἴη τὸ ἐμπύημα ἐπὶ θάτερα μοῦνον,[51] καταμανθάνειν χρὴ ἐπὶ τούτοισι, μή τι ἔχει ἄλγημα ἐν τῷ πλευρῷ· καὶ ἤν τι θερμότερον ᾖ τὸ ἕτερον τοῦ ἑτέρου· κατακλινομένου ἐπὶ τὸ ὑγιαῖνον πλευρὸν ἐρωτᾶν, εἴ τι αὐτῷ δοκέει βάρος ἐκκρέμασθαι ἐκ τοῦ ἄνωθεν. εἰ γὰρ εἴη τοῦτο, ἐκ τοῦ[52] ἐπὶ θάτερόν ἐστι τὸ ἐμπύημα, ἐπὶ ὁκοῖον ἂν πλευρὸν τὸ βάρος ἐγγίνηται.

17. Τοὺς δὲ ξύμπαντας ἐμπύους γινώσκειν χρὴ τοῖσδε τοῖσι σημείοισι· πρῶτον μὲν ὁ πυρετὸς οὐκ ἀφίησιν, ἀλλὰ τὴν μὲν ἡμέρην λεπτὸς ἴσχει, ἐς νύκτα

the ninth or the eleventh. Such are the conclusions to be drawn, since this sputum is a particularly deadly sign indicating that the patient will not reach the fourteenth day. The prognosis must be reckoned by weighing the favorable and unfavorable signs as they appear, since in this way you will be most likely to be correct. Most of the other suppurations have eruptions, some about the twentieth day, the thirtieth day, or the fortieth day, while yet others extend as far as the sixtieth day.

16. When a suppuration will begin must be calculated by reckoning from the day on which the patient first had fever, or when he was seized by a rigor, and whether he reported that the pain in the part of his body where he was suffering had turned into a heaviness. These then are the symptoms that occur at the beginning of an internal suppuration. Accordingly, the pus must be expected to break out after this time, in the periods I have indicated above. Whether the suppuration is on only one side must be learned from the following: whether or not there is pain in one side, and that side is warmer than the other one: lay the patient on his healthy side and ask him whether he feels some weight hanging down from above, for if this is so, the suppuration is (sc. only) on one side, the one from which the heaviness is coming.

17. All internal suppurations must be recognized by the following signs. First, the patient's fever does not remit, but it is lighter during the day and more violent at night.

---

51 Add. στρέφειν τε καὶ C'M

52 ἐκ τοῦ C'V: om. M

154 δὲ πλείων, καὶ ἱδρῶτες | πολλοὶ ἐπιγίνονται, βῆξαί τε θυμὸς ἐγγίνεται αὐτοῖσι, καὶ ἀποπτύουσιν οὐδὲν ἄξιον λόγου, καὶ οἱ μὲν ὀφθαλμοὶ ἔγκοιλοι γίνονται, αἱ δὲ γνάθοι ἐρυθήματα ἴσχουσιν, καὶ οἱ μὲν ὄνυχες τῶν χειρῶν γρυποῦνται, οἱ δὲ δάκτυλοι θερμαίνονται καὶ μάλιστα ἄκροι, καὶ ἐν τοῖσι ποσὶν οἰδήματα γίνεται, καὶ σιτίων οὐκ ἐπιθυμέουσι, καὶ φλύκταιναι γίνονται ἀνὰ τὸ σῶμα.

Ὁκόσα μὲν οὖν ἐγχρονίζει τῶν ἐμπυημάτων, ἴσχει τὰ σημεῖα ταῦτα καὶ πιστεύειν αὐτοῖσι χρὴ κάρτα· ὁκόσα δὲ ὀλιγοχρόνιά ἐστι, σημαίνεσθαι, τούτων ἤν τι ἐπιφαίνηται οἷα καὶ τοῖσιν ἐν ἀρχῇσι γινομένοισιν, ἅμα δὲ καὶ ἤν τι δυσπνούστερος ᾖ ὁ ἄνθρωπος. τὰ δὲ ταχύτερόν αὐτῶν καὶ βραδύτερον ῥηγνύμενα τοισίδε γινώσκειν χρή· ἢν μὲν ὁ πόνος ἐν ἀρχῇσι γένηται καὶ ἡ δύσπνοια καὶ ἡ βὴξ καὶ ὁ πτυελισμὸς διατελέῃ[53] ἔχων, ἐς τὰς εἴκοσιν ἡμέρας προσδέχεσθαι τὴν ῥῆξιν ἢ καὶ ἔτι πρόσθεν· ἢν δὲ ἡσυχέστερος ὁ πόνος ᾖ καὶ τὰ ἄλλα πάντα κατὰ λόγον, τούτοισι προσδέχεσθαι

156 τὴν | ῥῆξιν ὕστερον· προγενέσθαι δὲ ἀνάγκη καὶ πόνον καὶ δύσπνοιαν καὶ πτυελισμὸν πρὸ τῆς τοῦ πύου ῥήξιος.

Περιγίνονται δὲ τούτων μάλιστα μὲν οὓς ἂν ἀφῇ ὁ πυρετὸς αὐθημερὸν μετὰ τὴν ῥῆξιν καὶ σιτίων ταχέως ἐπιθυμέωσι καὶ δίψης ἀπηλλαγμένοι ἔωσι καὶ ἡ γαστὴρ σμικρά τε καὶ ξυνεστηκότα ὑποχωρέῃ καὶ τὸ πύον λευκόν τε καὶ λεῖον καὶ ὁμόχροον ᾖ[54] καὶ φλέγματος ἀπηλλαγμένον καὶ ἄτερ πόνου τε καὶ βηχὸς

Abundant sweats follow; but although in such cases there is a strong desire to cough, patients do not cough up anything worth mentioning. The eyes are hollow, the cheeks flushed, and the fingernails clubbed; the fingers are warm, especially at their tips, swelling develops in the feet, there is no inclination to eat, and blisters cover the body.

Now internal suppurations that have been present for a longer time present the following signs, in which you must place much trust, while in suppurations that have been present for only a short time, you must make your conjecture on the basis of any sign that appears, from the signs that were present at the beginning, and if at the same time the patient has a degree of dyspnea. You must learn whether the rupture (sc. of the suppuration) will occur more quickly or later, by the following signs: if pain is present from the beginning, and the dyspnea, cough, and expectoration are continuous, expect the rupture in twenty days, or even before that. But if the pain is more leisurely, and all the other symptoms quite similar, expect the rupture to be later. In any case, before the rupture of the pus, there will certainly be pain, dyspnea and expectoration.

The patients with suppurations who are most likely to survive are the ones in whom the fever disappears after the rupture on the same day, who soon desire food, who are relieved of their thirst, whose cavity passes scanty, compact stools, and whose pus runs out white, smooth, uniform in color, and devoid of phlegm, and who are cleaned without pain or violent coughing: these recover

---

[53] διατελέῃ Jouanna: διατελέει C′: διατείνῃ MV
[54] ᾗ MV: ἐκχωρέει C′

ἰσχυρῆς ἀνακαθαίρηται. ἄριστα μὲν οὖν οὗτοι[55] καὶ τάχιστα ἀπαλλάσσουσιν· εἰ δὲ μή, οἷσιν ἂν ἐγγυτάτω τούτων γίνηται. ἀπόλλυνται δὲ οὓς ἂν ὁ πυρετὸς αὐθημερὸν μὴ ἀφῇ ἢ[56] δοκέων ἀφιέναι αὖθις φαίνηται ἀναθερμαινόμενος, καὶ δίψαν μὲν ἔχωσι, σιτίων δὲ μὴ ἐπιθυμέωσι, καὶ ἡ κοιλίη ὑγρὴ ᾖ καὶ τὸ πύον χλωρὸν καὶ πελιὸν πτύῃ ἢ φλεγματῶδες καὶ ἀφρῶδες· οἷσιν ἂν ταῦτα πάντα γίνεται, ἀπόλλυνται· ὁκόσοισι δ' ἂν τούτων τὰ μὲν ἐπιγένηται, τὰ δὲ μή, οἱ μὲν αὐτῶν ἀπόλλυνται, οἱ δὲ ἐν πολλῷ χρόνῳ περιγίνονται. | 158 ἀλλ' ἐκ πάντων τῶν τεκμηρίων τῶν ἐόντων ἐν τούτοισι σημαίνεσθαι[57] καὶ τοῖσιν ἄλλοισιν πᾶσιν.

18. Ὁκόσοισι δὲ ἀποστάσιες γίνονται ἐκ τῶν περιπνευμονικῶν νοσημάτων[58] παρὰ τὰ ὦτα καὶ ἐκπυέουσιν, ἢ[59] ἐς τὰ κάτω χωρία καὶ συριγγοῦνται, οὗτοι δὲ περιγίνονται. ὑποσκέπτεσθαι δὲ χρὴ τὰ τοιαῦτα ὧδε· ἢν ὅ τε πυρετὸς ἔχῃ καὶ ἡ ὀδύνη μὴ πεπαυμένη ᾖ καὶ τὸ πτύελον μὴ ἐκχωρέῃ κατὰ λόγον, μηδὲ χολώδεες αἱ τῆς κοιλίης διαχωρήσιες μηδὲ εὔλυτοί τε καὶ ἄκρητοι[60] γίνωνται, μηδὲ τὸ οὖρον πολύ[61] τε κάρτα καὶ 160 πολλὴν | ὑπόστασιν ἔχον, ὑπηρετῆται δὲ περιεστικῶς ὑπὸ τῶν λοιπῶν πάντων τῶν περιεστικῶν σημείων, τούτοισι χρὴ τὰς τοιαύτας ἀποστάσιας ἐλπίζειν ἔσεσθαι. γίνονται δὲ αἱ μὲν ἐς τὰ κάτω χωρία, οἷσιν ἄν τι περὶ τὰ ὑποχόνδρια τοῦ φλέγματος ἐγγίνηται, αἱ δὲ ἄνω, οἷσιν ἂν τὸ μὲν ὑποχόνδριον λαπαρόν τε καὶ ἀνώδυνον διατελέῃ ἐόν, δύσπνοος δέ τινα χρόνον γενόμενος παύσηται ἄτερ φανερῆς προφάσιος ἄλλης.

best and most quickly; otherwise, patients who are most similar to these. Patients die in whom the fever does not subside on the same day (sc. as the rupture) or while seeming to abate reappears and warms them up again, who have thirst, who do not desire food, whose cavity is fluent, and who cough up pus that is green and livid or phlegmy and foamy. Those with all these signs die, while those with some but not others sometimes die, but in other cases recover after a long time. Judgments regarding these patients must be made on the basis of all the signs present, just as in other cases.

18. Patients in whom, after pneumonic diseases, abscessions form beside the ears and suppurate, or in the lower parts (sc. of the chest) with fistulas, survive. Investigate such cases thus: if fever is present, the pain has not ceased, sputum has not been properly coughed up, the stools evacuated from the cavity are not bilious, fluent, and well-compounded, and the urine is scanty and lacking much sediment, but the patients do benefit from all the other favorable signs, there is reason to expect the formation of such abscesses. Some of these form in the lower parts (sc. of the chest), when the phlegm invades the hypochondrium, and others above this with the patient's hypochondrium remaining relaxed and pain-free, while he experiences dyspnea for a certain time, but then recovers for no other apparent reason.

---

55 *οὗτοι* G′: *οὕτως* M: *οὕτω* C′V 56 *ἢ* Jouanna: *ἀλλὰ* C′: *καὶ* MV 57 *σημ.* MV: *τεκμαίρεσθαι* C′ 58 *νοσημάτων* C′: om. MV 59 *ἢ* C′: om. MV 60 *ἄκρητοι* M: *εὔκριτοι* C′: *ἄκριτοι* V 61 *πολύ* MV: *παχύ* C′

Αἱ δὲ ἀποστάσιες αἱ ἐς τὰ σκέλεα ἐν τῇσι περιπλευμονίῃσι τῇσιν ἰσχυρῇσι καὶ ἐπικινδύνοισι λυσιτελέες μὲν πᾶσαι, ἄρισται δὲ αἱ τοῦ πτυέλου ἐν μεταβολῇ ἐόντος ἤδη γινόμεναι· εἰ γὰρ τὸ οἴδημα καὶ ἡ ὀδύνη γίνοιτο, τοῦ πτυέλου ἀντὶ ξανθοῦ πυώδεος γινομένου καὶ ἐκχωρέοντος ἔξω, οὕτως ἂν ἀσφαλέστατα ὅ τε ἄνθρωπος περιγίνοιτο, καὶ ἡ ἀπόστασις τάχιστα ἀνώδυνος ἂν παύσαιτο· εἰ δὲ τὸ πτύελον μὴ ἐκχωρέοι καλῶς, μηδὲ τὸ οὖρον ὑπόστασιν ἀγαθὴν ἔχον φαίνοιτο, κίνδυνος γενέσθαι χωλὸν τὸ | ἄρθρον ἢ πολλὰ πρήγματα παρασχεῖν. ἢν δὲ ἀφανίζωνται καὶ παλινδρομέωσιν αἱ ἀποστάσιες τοῦ πτυέλου μὴ ἐκχωρέοντος τοῦ τε πυρετοῦ ἔχοντος, δεινόν· κίνδυνος γὰρ μὴ παραφρονήσῃ καὶ ἀποθάνῃ ὁ ἄνθρωπος. τῶν δὲ ἐμπύων τῶν ἐκ τῶν περιπνευμονικῶν οἱ γεραίτεροι μᾶλλον ἀπόλλυνται· ἐκ δὲ τῶν ἄλλων ἐμπυημάτων οἱ νεώτεροι μᾶλλον ἀποθνῄσκουσιν. |

19. Αἱ δὲ ξὺν πυρετῷ ὀδύναι γινόμεναι περὶ τὴν ὀσφύν τε καὶ τὰ κάτω χωρία, ἢν τῶν φρενῶν ἅπτωνται, τὰ κάτω ἐκλείπουσαι, ὀλέθριαι κάρτα. προσέχειν οὖν δεῖ τὸν νόον τοῖσιν ἄλλοισι σημείοισιν, ὡς ἤν τι καὶ τῶν ἄλλων σημείων πονηρὸν ἐπιφαίνηται, ἀνέλπιστος ὁ ἄνθρωπος· ἢν δὲ ἀναΐσσοντος τοῦ νοσήματος [ὡς][62] πρὸς τὰς φρένας τἆλλα σημεῖα μὴ πονηρὰ ἐπιγίνηται, ἔμπυον ἔσεσθαι πολλαὶ ἐλπίδες τοῦτον. |

Κύστιες δὲ σκληραί τε καὶ ἐπώδυνοι δειναὶ μὲν παντελῶς καὶ ὀλέθριοι,[63] ὀλεθριώταται δὲ ὁκόσαι ξὺν πυρετῷ συνεχεῖ γίνονται· καὶ γὰρ οἱ ἀπ᾽ αὐτῶν τῶν

Abscessions migrating to the legs in severe and dangerous pneumonias are all beneficial, and best are those doing this while the sputum is in the process of changing. For if swelling and pain set in, simultaneously with the sputum turning from yellow to purulent and being expelled, the person will most assuredly survive, and the abscession will very soon end without pain. However, if the sputum is not evacuated properly, and the urine fails to contain a favorable sediment, there is a danger that some joint will become lame or cause many other difficulties. If the abscessions recede and disappear without the expectoration of any sputum, and with fever still present, this is ominous, since there is a danger that the patient will lose his mind and die. From internal suppurations following pneumonia older people are more likely to die, but from other internal suppurations younger ones are more likely to die.

19. Pains accompanied by fever which arise around the loins and lower parts, if they move away from the lower parts and invade the diaphragm are very dire. Attention must be directed to other signs as well, since if any of these is unfavorable, the person has no hope of survival; however if, when the condition shoots up against the diaphragm, the patient has other signs that are not too bad, the chance is great that he will have an internal suppuration.

Bladders being hard and painful are unfavorable in every way and fatal, but most deadly if accompanied by a continuous fever. In fact, pains coming from the bladder

---

[62] ὡς codd.: del. Jones

[63] παντελῶς καὶ ὀλέθριοι M: πᾶσαι C'V

κυστίων πόνοι ἱκανοὶ ἀποκτεῖναι, καὶ αἱ κοιλίαι οὐ διαχωρέουσιν ἐν τούτῳ τῷ χρόνῳ,[64] *εἰ μὴ σκληρά τε καὶ πρὸς ἀνάγκην. λύει δὲ οὖρον πυῶδες οὐρηθέν,*
168 *λευκὴν καὶ λείην ἔχον τὴν ὑπόστασιν· ἢν δὲ μήτε | τὸ οὖρον μηδὲν ἐνδιδῷ μήτε ἡ κύστις μαλθάσσηται ὅ τε πυρετὸς ξυνεχὴς ᾖ, ἐν τῇσι πρώτῃσι περιόδοισι τοῦ νοσήματος ἐλπὶς τὸν ἀλγέοντα ἀπολέσθαι· ὁ δὲ τρόπος οὗτος μάλιστα τῶν παιδίων ἅπτεται τῶν ἀπὸ ἑπτὰ ἐτέων, ἔστ' ἂν πεντεκαιδεκαετέες γένωνται.*

20. *Οἱ δὲ πυρετοὶ κρίνονται ἐν τῇσιν αὐτῇσιν ἡμέρῃσι τὸν ἀριθμόν, ἐξ ὧν τε περιγίνονται οἱ ἄνθρωποι καὶ ἐξ ὧν ἀπόλλυνται. οἵ τε γὰρ εὐηθέστατοι τῶν πυρετῶν καὶ ἐπὶ σημείων ἀσφαλεστάτων βεβῶτες τεταρταῖοι παύονται ἢ πρόσθεν, οἵ τε κακοηθέστατοι καὶ ἐπὶ σημείων δεινοτάτων γινόμενοι τεταρταῖοι κτείνουσιν ἢ πρόσθεν. ἡ μὲν οὖν πρώτη ἔφοδος αὐτῶν οὕτω τελευτᾷ· ἡ δὲ δευτέρη ἐς τὴν ἑβδόμην περιάγεται, ἡ δὲ τρίτη ἐς τὴν ἑνδεκάτην, ἡ δὲ τετάρτη ἐς τὴν τεσσαρεσκαιδεκάτην, ἡ δὲ πέμπτη ἐς τὴν ἑπτακαιδεκάτην, ἡ δὲ ἕκτη ἐς τὴν εἰκοστήν. αὗται μὲν οὖν ἐκ τῶν ὀξυτάτων νοσημάτων διὰ τεσσάρων ἐς τὰς εἴκοσιν ἐκ προσθέσιος τελευτῶσιν. οὐ δύναται δὲ ὅλησιν*
170 *ἡμέρῃσιν οὐδὲν τούτων | ἀριθμεῖσθαι ἀτρεκέως· οὐδὲ γὰρ ὁ ἐνιαυτός τε καὶ οἱ μῆνες ὅλῃσιν ἡμέρῃσι πεφύκασιν ἀριθμεῖσθαι.*

Μετὰ δὲ ταῦτα ἐν τῷ αὐτῷ τρόπῳ κατὰ τὴν αὐτὴν πρόσθεσιν ἡ μὲν πρώτη περίοδος τεσσάρων καὶ τριήκοντα ἡμερέων, ἡ δὲ δευτέρη τεσσαράκοντα ἡμερέων,

are sufficient in themselves to cause death, and furthermore cavities do not evacuate at that time, except with hard stools and under compulsion. The passage of purulent urine containing a fine, white sediment will resolve this. But if the urine does not relent at all, and the bladder does not become softer and a continuous fever remains, it is to be expected that the sufferer will succumb during the first cycles of the disease. This kind of disorder seizes mainly children between seven and fifteen years.

20. Fevers have their crisis in the same number of days, both those from which patients recover, and those from which they die. The most harmless fevers with the most favorable signs end on the fourth day or before, just as the most malignant fevers with the most unfavorable signs cause death on the fourth day or before. This is how the first access ends, while the second reaches the seventh day, the third the eleventh day, the fourth the fourteenth day, the fifth the seventeenth day, and the sixth the twentieth day: that is, fevers from the most acute diseases end, by the addition of four-day periods, at twenty. But it is not possible to calculate these exactly in whole days, for after all neither the year nor the months are such that they can be calculated in whole days.

Beyond this, making the same addition in the same way, a first period will be thirty-four days, a second forty,

---

64 *ἐν τ. τ. χρόνῳ* MV: *ἐπὶ τῶν τοιούτων* C′

ἡ δὲ τρίτη ἑξήκοντα ἡμερέων. τούτων δ' ἐν ἀρχῇσίν ἐστι χαλεπώτατα προγινώσκειν τὰ μέλλοντα ἐν πλείστῳ χρόνῳ κρίνεσθαι· ὁμοιόταται γὰρ αἱ ἀρχαὶ εἰσιν αὐτῶν· ἀλλὰ χρὴ ἀπὸ τῆς πρώτης ἡμέρης ἐνθυμέεσθαι καὶ καθ' ἑκάστην τετράδα προστιθεμένην σκέπτεσθαι καὶ οὐ λήσει, ὅπῃ τρέψεται. γίνεται δὲ καὶ τῶν τεταρταίων ἡ κατάστασις ἐκ τούτου τοῦ κόσμου. τὰ δὲ ἐν ἐλαχίστῳ χρόνῳ μέλλοντα κρίνεσθαι εὐπετέστερα γινώσκεσθαι· μέγιστα γὰρ τὰ διαφέροντα αὐτῶν ἐστιν ἀπ' ἀρχῆς· οἱ μὲν γὰρ περιεσόμενοι εὔπνοοί τε καὶ ἀνώδυνοί εἰσι καὶ κοιμῶνται τὰς νύκτας τά τε ἄλλα σημεῖα ἔχουσιν ἀσφαλέστατα· οἱ δὲ ἀπολλύμενοι δύσπνοοι γίνονται, ἀλλοφάσσοντες, ἀγρυπνέοντες τά τε ἄλλα σημεῖα κάκιστα ἔχοντες. ὡς
172 οὖν τούτων οὕτω | γινομένων ξυμβάλλεσθαι χρὴ κατά τε τὸν χρόνον κατά τε τὴν πρόσθεσιν ἑκάστην ἐπὶ τὴν κρίσιν ἰόντων τῶν νοσημάτων. κατὰ δὲ τὸν αὐτὸν τρόπον καὶ τῇσι γυναιξὶν αἱ κρίσιες ἐκ τῶν τόκων γίνονται.

21. Κεφαλῆς δὲ ὀδύναι ἰσχυραί τε καὶ ξυνεχέες ξὺν πυρετῷ, εἰ μέν τι τῶν θανατωδέων σημείων προσγίνοιτο, ὀλέθριον κάρτα· εἰ δὲ ἄτερ σημείων τοιούτων ἡ ὀδύνη ὑπερβάλλοι εἴκοσιν ἡμέρας ὅ τε πυρετὸς ἔχοι, ὑποσκέπτεσθαι χρὴ αἵματος ῥῆξιν <ἢ ἐκπύησιν>[65] διὰ ῥινῶν ἢ ἄλλην ἀπόστασιν ἐς τὰ κάτω χωρία. ἔστ' ἂν δὲ ἡ ὀδύνη ᾖ νεαρά, προσδέχεσθαι χρὴ αἵματος ῥῆξιν ἢ ἐκπύησιν διὰ ῥινῶν, ἄλλως τε κἢν ἡ ὀδύνη περὶ τοὺς κροτάφους τε καὶ τὸ μέτωπον ᾖ. μᾶλλον δὲ χρὴ

and a third sixty days. At the beginning of these periods it is very difficult to know in advance what crises will occur over a very long time, since their onsets all resemble each other. But you must consider the matter, starting from the first day, and examine them each time four days are added, for in this way it will not escape you where the fever is heading. The scheme of quartan fevers is also according to this same order. But fevers that are going to have their crisis in a shorter time are easier to understand, since the greatest differences are in evidence from their beginning. Patients who are going to survive breathe easily and are free of pain, they sleep at night, and display the rest of the safest signs, while those who are destined to die suffer from dyspnea, delirium, and sleeplessness, and have the other most dangerous signs. Now since these things happen like this, you must consider them through time and by each added step toward the diseases' crisis as they advance. In women the crises occurring after giving birth occur in the same way as well.

21. Severe continuous headaches accompanied by fever are very ominous if any of the fatal signs should be added. But if without the appearance of any such signs the pain and fever persist beyond twenty days, you must expect there to be a hemorrhage <or expulsion of pus> from the nostrils, or some different abscession in the lower parts. While the pain is still recent, expect hemorrhaging through the nostrils or an expulsion of pus, especially if the pain is located around the temples and the forehead.

---

[65] Add. Jouanna, cf. *Coan Prenotions* 156

προσδέχεσθαι τοῦ μὲν αἵματος τὴν ῥῆξιν τοῖσι νεωτέροισι πέντε καὶ τριήκοντα ἐτέων, τοῖσι δὲ γεραιτέροισι τὴν ἐκπύησιν. |

174 22. Ὠτὸς δὲ ὀδύνη ὀξεῖα ξὺν πυρετῷ ξυνεχεῖ τε καὶ ἰσχυρῷ δεινόν· παραφρονῆσαι γὰρ κίνδυνος τὸν ἄνθρωπον καὶ ἀπολέσθαι. ὡς οὖν τούτου τοῦ τρόπου σφαλεροῦ ἐόντος ταχέως[66] δεῖ προσέχειν τὸν νόον καὶ τοῖσιν ἄλλοισι σημείοισι πᾶσιν ἀπὸ τῆς πρώτης ἡμέρης. ἀπόλλυνται δὲ οἱ μὲν νεώτεροι τῶν ἀνθρώπων ἑβδομαῖοι καὶ ἔτι θᾶσσον ὑπὸ τοῦ νοσήματος τούτου, οἱ δὲ γέροντες πολλῷ βραδύτερον· οἵ τε γὰρ πυρετοὶ καὶ αἱ παραφροσύναι ἧσσον αὐτοῖσιν ἐπιγίνονται, καὶ τὰ ὦτα αὐτοῖσι διὰ τοῦτο φθάνει ἐκπυούμενα· ἀλλὰ ταύτῃσι μὲν τῇσιν ἡλικίῃσιν ὑποστροφαὶ τοῦ νοσήματος ἐπιγινόμεναι ἀποκτείνουσι τοὺς πλείστους· οἱ δὲ νεώτεροι, πρὶν ἐκπυῆσαι τὸ οὖς, ἀπόλλυνται. ἐπεὶ ἤν[67] γε ῥυῇ πύον λευκὸν ἐκ τοῦ ὠτός, ἐλπὶς περιγενέσθαι τῷ νέῳ, ἤν τι καὶ ἄλλο χρηστὸν αὐτῷ ἐπιγένηται σημεῖον.

23. Φάρυγξ δὲ ἑλκουμένη ξὺν πυρετῷ δεινόν· ἀλλ' 176 ἤν τι | καὶ ἄλλο σημεῖον ἐπιγένηται τῶν προκεκριμένων πονηρῶν εἶναι, προλέγειν ὡς ἐν κινδύνῳ ἐόντος τοῦ ἀνθρώπου. αἱ δὲ κυνάγχαι δεινόταται μέν εἰσι καὶ τάχιστα ἀναιρέουσιν, ὁκόσαι μήτ' ἐν τῇ φάρυγγι μηδὲν ἔκδηλον ποιέουσι μήτ' ἐν τῷ αὐχένι, πλεῖστόν τε πόνον παρέχουσι καὶ ὀρθόπνοιαν· αὗται γὰρ καὶ αὐθημερὸν ἀποπνίγουσι καὶ δευτεραῖον καὶ τριταῖον καὶ τεταρταῖον. ὁκόσαι δὲ τὰ μὲν ἄλλα παραπλησίως

Hemorrhage is more to be expected in people younger than thirty-five years, suppuration in those who are older.

22. Acute pain in an ear accompanied by a continuous, sharp fever is serious; for it indicates the danger that such patients will become delirious and die. Since this condition is treacherous, immediate attention must be paid to all the other signs too, from the first day on. Younger people die from this condition on the seventh day or even more quickly, but older ones much more slowly, since the fever and delirium affect them less, and for this reason their ears erupt before they can die. But if at these (sc. higher) ages the disease relapses, it is fatal for the majority of patients, whereas younger patients die before their ear ever suppurates. If white pus is expelled from the ear of a younger person, there is a hope of survival if he has some other favorable sign.

23. Ulceration of the throat in conjunction with fever is serious, but if some other sign among those that have already been determined to be dangerous is added, announce that the patient is in danger. Anginas are most ominous and rapidly lethal if they lack any visible effect in the throat or neck, but cause much suffering as well as orthopnea; these can suffocate the patient as early as the first day, but also on the second, third, and fourth day. Cases like this that also provoke pain, and cause swelling

---

66 ταχέως MV: ὀξέως C′
67 ἐπεὶ ἢ̓ν V: ἐπὴν C′M

πόνον παρέχουσιν, ἐπαίρονται δὲ καὶ ἐρυθήματα ἐν τῇ φάρυγγι ποιέουσιν, ὀλέθριαι μὲν κάρτα, χρονιώτεραι δὲ μᾶλλον τῶν πρόσθεν. ὁκόσοισι δὲ ξυνεξερεύθει ἡ φάρυγξ καὶ ὁ αὐχήν, αὗται δὲ χρονιώτεραι, καὶ
178 μάλιστα ἐξ αὐτῶν περιγίνονται, ἢν ὅ τε | αὐχὴν καὶ τὸ στῆθος ἐρυθήματα ἔχωσιν καὶ μὴ παλινδρομέῃ τὸ ἐρυσίπελας ἔσω. ἢν δὲ μήτε ἐν ἡμέρῃσι κρισίμῃσιν ἀφανίζηται τὸ ἐρυσίπελας, μήτε φύματος ξυστραφέντος ἐν τῷ ἔξω χωρίῳ, μήτε πῦον ἀποβήσσῃ ῥηϊδίως τε καὶ ἀπόνως, θάνατον σημαίνει ἢ ὑποστροφὴν τοῦ ἐρυθήματος. ἀσφαλέστατον δὲ τὸ ἐρύθημα καὶ τὸ οἴδημα ὡς μάλιστα ἔξω τρέπεσθαι·[68] ἢν δὲ ἐς τὸν πνεύμονα τρέπηται, παράνοιάν[69] τε ποιεῖ καὶ ἔμπυοι ἐξ αὐτῶν γίνονται ὡς τὰ πολλά.

Οἱ δὲ γαργαρεῶνες ἐπικίνδυνοι καὶ ἀποτάμνεσθαι καὶ ἀποσχάζεσθαι, ἔστ᾽ ἂν ἐρυθροί τε ἔωσι καὶ μεγάλοι· καὶ γὰρ φλεγμοναὶ ἐπιγίνονται τούτοισι καὶ αἱμορραγίαι· ἀλλὰ χρὴ τὰ τοιαῦτα τοῖσιν ἄλλοισι μηχανήμασι πειρῆσθαι κατισχναίνειν ἐν τούτῳ τῷ χρόνῳ. ὁκόταν δὲ ἀποκριθῇ ἤδη,[70] ὃ δὴ σταφυλὴν
180 καλέουσι, καὶ γένηται τὸ | μὲν ἄκρον τοῦ γαργαρεῶνος μεῖζόν τε καὶ περιφερές,[71] τὸ δὲ ἀνωτέρω λεπτότερον, ἐν τούτῳ τῷ καιρῷ ἀσφαλὲς διαχειρίζειν. ἄμεινον δὲ καὶ ὑποκενώσαντα τὴν κοιλίην τῇ χειρουργίῃ

[68] Add. καὶ τὰς ἑτέρας ἀποστάσιας ἔξω τρέπεσθαι C′
[69] C′ has a gap in its text from -νοίαν to ch. 24 γεραιτέροισιν.
[70] Add. πᾶν V
[71] περιφερές M: πελιόν V

and redness in the throat are definitely sinister, although they continue longer than the ones already mentioned. If their throat and neck become red simultaneously, patients are ill for a longer time, but usually recover if their neck and chest have erythemas without the erysipelas migrating to the interior. But if the erysipelas does not disappear on the critical days, if no growth forms on the outside, and if the patient does not expectorate pus easily and painlessly, this announces death or a return of the erythema. It is safest for erythema and swelling to turn outward as much as possible. But if they turn into the lung, they cause delirium, and in most cases the patients suppurate internally from them.

Cases of staphylitis[4] are dangerous to excise or scarify as long as they are red and swollen, for both inflammation and hemorrhaging may result. Rather, you should attempt to reduce such swellings at this time by other means. When what they call the "grape" has already been formed, and the tip of the uvula is enlarged and round, but the stalk higher up is thinner, then it is safe to operate. It is also better if you empty the cavity downward be-

[4] In the Hippocratic Collection the terms γαργαρέων and σταφυλή are used inconsistently for the uvula in its normal and inflamed state (staphylitis): in *Affections* 4 σταφυλή is the anatomical term and γαργαρέων the pathological, whereas in *Diseases II* 10 it is the opposite. Here in *Prognostic* the usage is not clear: at first, γαργαρέων is the disease, but later, σταφυλή too seems to be the pathological state. See Dönt, pp. 78–80; Skoda, pp. 247–49; Jouanna *Prog.*, p. 69, n. 1.

χρέεσθαι, ἢν ὅ τε χρόνος ξυγχωρέῃ καὶ μὴ ἀποπνίγηται ὁ ἄνθρωπος.

24. Ὁκόσοισι δ' ἂν οἱ πυρετοὶ παύωνται μήτε σημείων γενομένων λυτηρίων μήτε ἐν ἡμέρῃσι κρισίμῃσιν, ὑποστροφὴν προσδέχεσθαι τούτοισιν. ὅστις δ' ἂν τῶν πυρετῶν μηκύνῃ περιεστικῶς διακειμένου τοῦ ἀνθρώπου, μήτε ὀδύνης ἐχούσης διὰ φλεγμονήν τινα μήτε διὰ πρόφασιν ἄλλην μηδεμίαν ἐμφανέα, τούτῳ προσδέχεσθαι ἀπόστασιν μετ' οἰδήματός τε καὶ ὀδύνης ἔς τι τῶν ἄρθρων καὶ οὐχ ἧσσον τῶν κάτω. μᾶλλον δὲ γίνονται καὶ ἐν ἐλάσσονι χρόνῳ αἱ τοιαῦται ἀποστάσιες τοῖσι νεωτέροισι τριήκοντα ἐτέων. ὑποσκέπτεσθαι δὲ χρὴ εὐθέως τὰ περὶ τῆς ἀποστάσιος, ἢν εἴκοσιν ἡμέρας ὁ πυρετὸς ἔχων ὑπερβάλλῃ. τοῖσι 182 δὲ γεραιτέροισιν | ἧσσον γίνεται πολυχρονιωτέρου ἐόντος τοῦ πυρετοῦ. χρὴ δὲ τὴν μὲν τοιαύτην ἀπόστασιν προσδέχεσθαι ξυνεχέος ἐόντος τοῦ πυρετοῦ, ἐς δὲ τεταρταῖον καταστήσεσθαι, ἢν διαλείπῃ τε καὶ καταλαμβάνῃ πεπλανημένον τρόπον καὶ ταῦτα ποιέων τῷ φθινοπώρῳ πελάσῃ. ὥσπερ δὲ τοῖσι νεωτέροισι τριήκοντα ἐτέων αἱ ἀποστάσιες γίνονται, οὕτως οἱ τεταρταῖοι μᾶλλον τοῖσι τριήκοντα ἐτέων καὶ γεραιτέροισιν. τὰς δὲ ἀποστάσιας εἰδέναι χρὴ τοῦ χειμῶνος μᾶλλον γινομένας χρονιώτερόν τε παυομένας, ἧσσον δὲ παλινδρομεύσας.

Ὅστις δ' ἂν ἐν πυρετῷ μὴ θανατώδει φῇ τὴν κεφαλὴν ἀλγέειν ἢ καὶ ὀρφνῶδές τι πρὸ τῶν ὀφθαλμῶν γίνεσθαι, ἢν καρδιωγμὸς τούτῳ προσγένηται, χο-

fore the surgery, if time allows it and the person is not suffocating.

24. If fevers end without the usual salutary signs, and not on the critical days, expect a relapse in such patients. Any of the fevers that continues longer, although the patient is in a condition indicating recovery, suffering no pain due to inflammation or any other evident cause, you may expect to lead to an abscession together with edema, with the patient suffering pains in some joint, not least often a lower one. Such abscessions are more likely to occur and in a shorter time, in people who are younger than thirty years. You must immediately suspect that there is an abscession if such a fever continues for more than twenty days. In older patients the abscession occurs less often and after the fever has been present for a longer time. You must expect, in this kind of abscession, if the fever is continuous but then intermits and attacks in an erratic pattern, for it to settle into a quartan, and to do this with the approach of fall. Just as the abscessions tend to occur in persons less than thirty years, the quartan fevers are more frequent in those over thirty years. Be clear that abscessions in winter occur more frequently and last longer, but that they relapse less often.

A person with a nonfatal fever who reports that he has a headache or also that something dark is appearing before his eyes, if befallen in addition by heartburn, will bring up

λώδης ἔμετος παρέσται· ἢν δὲ καὶ ῥῖγος προσγένηται καὶ τὰ κάτω τοῦ ὑποχονδρίου ψυχρὰ ἔχῃ, καὶ θᾶσσον ἔτι ὁ ἔμετος παρέσται· ἢν δέ τι πίῃ ἢ φάγῃ ὑπὸ
184 τοῦτον τὸν χρόνον, κάρτα | ταχέως ἐμεῖται. τούτων δὲ οἷσιν ἂν ἄρξηται ὁ πόνος τῇ πρώτῃ ἡμέρῃ γίνεσθαι, τεταρταῖοι πιεζεῦνται μάλιστα καὶ πεμπταῖοι· ἐς δὲ τὴν ἑβδόμην ἀπαλλάσσονται· οἱ μέντοι πλεῖστοι αὐτῶν ἄρχονται μὲν πονέεσθαι τριταῖοι, χειμάζονται δὲ μάλιστα πεμπταῖοι· ἀπαλλάσσονται δὲ ἐναταῖοι ἢ ἑνδεκαταῖοι· οἳ δ' ἂν ἄρξωνται πεμπταῖοι πονέεσθαι καὶ τὰ ἄλλα κατὰ λόγον αὐτοῖσι τῶν πρόσθεν γίνηται, ἐς τὴν τεσσαρεσκαιδεκάτην κρίνεται ἡ νοῦσος. γίνεται δὲ ταῦτα τοῖσι μὲν ἀνδράσι καὶ τῇσι γυναιξὶν ἐν τοῖσι τριταίοισι μάλιστα· τοῖσι δὲ νεωτέροισι γίνεται μὲν καὶ ἐν τούτοισι, μᾶλλον δὲ ἐν τοῖσι ξυνεχεστέροισι πυρετοῖσι καὶ ἐν τοῖσι γνησίοισι τριταίοισιν.

Οἷσι δ' ἂν ἐν τοιουτοτρόπῳ πυρετῷ κεφαλὴν ἀλγέουσιν ἀντὶ μὲν τοῦ ὀρφνῶδές τι πρὸ τῶν ὀφθαλμῶν φαίνεσθαι ἀμβλυωγμὸς γίνηται ἢ μαρμαρυγαὶ προφαίνωνται, ἀντὶ δὲ τοῦ καρδιώσσειν ἐν τῷ ὑποχον-
186 δρίῳ ἐπὶ | δεξιὰ ἢ ἐπ' ἀριστερὰ ξυντείνηταί τι μήτε ξὺν ὀδύνῃ μήτε ξὺν φλεγμονῇ, αἷμα διὰ ῥινῶν ῥαγῆναι τούτοισι προσδόκιμον ἀντὶ τοῦ ἐμέτου. μᾶλλον δὲ καὶ ἐνταῦθα τοῖσι νέοισι τοῦ αἵματος τὴν ῥῆξιν προσδέχεσθαι· τοῖσι δὲ τριηκονταέτεσι καὶ γεραιτέροισιν ἧσσον,[72] ἀλλὰ τοὺς ἐμέτους τούτοισι προσδέχεσθαι.

Τοῖσι δὲ παιδίοισι σπασμοὶ γίνονται, ἢν ὅ τε πυρετὸς ὀξὺς ᾖ καὶ ἡ γαστὴρ μὴ διαχωρέῃ καὶ ἀγρυ-

bilious vomitus. If in addition a chill comes on, and the parts below his hypochondrium are cold, the vomiting will occur even sooner. Or if he drinks or eats anything at this time, he will vomit very soon. In patients whose pain began on the first day, the distress is greatest on the fourth and fifth day, and relief follows toward the seventh day. Most, however, first suffer pain on the third day, are most severely afflicted on the fifth day, and obtain relief on the ninth or eleventh day. When the pains begin on the fifth day, and the signs follow in the same order as in the previous cases, then the disease has its crisis toward the fourteenth day. In men and women such a course is most often in a tertian fever, while in younger people, although it may also be in a tertian fever, it is more frequent in more continuous fevers and in genuine tertians.[5]

Patients who have this kind of fever together with headache, but who instead of seeing something dark before their eyes have dimness of vision and see sparks, and rather than having heartburn feel a tension in the hypochondrium on the right or the left without pain or inflammation, you may expect to hemorrhage from the nostrils rather than to vomit. Here too expect the epistaxis more frequently in the young, but less often in those over thirty years, in whom you should rather expect the vomiting.

In children convulsions occur if their fever is acute, if their belly passes nothing, if they do not sleep, if they are

[5] See Jones, vol. 2, p. 53, n. 1: "*I.e.* tertians that *intermit*, the fever ceasing entirely every other day. Many tertians *remit* only, the fever growing less instead of ceasing altogether."

---

[72] The text recommences in C′ with ἦσσον.

πνέωσί τε καὶ ἐκπλαγέωσι καὶ κλαυθμυρίζωσι καὶ τὸ χρῶμα μεταβάλλωσι καὶ χλωρὸν ἢ πελιὸν ἢ ἐρυθρὸν ἴσχωσι. γίνεται δὲ ταῦτα ἐξ ἑτοιμοτάτου μὲν τοῖσι παιδίοισι τοῖσι νεωτάτοισιν ἐς τὰ ἑπτὰ ἔτεα· τὰ δὲ πρεσβύτερα τῶν παιδίων καὶ οἱ ἄνδρες οὐκ ἔτι ἐν τοῖσι πυρετοῖσιν ὑπὸ τῶν σπασμῶν ἁλίσκονται, ἢν μή τι τῶν σημείων προσγένηται τῶν ἰσχυροτάτων τε
188 καὶ κακίστων, οἷά περ ἐπὶ | τῇσι φρενίτισι γίνεται. τοὺς δὲ περιεσομένους τε καὶ ἀπολουμένους τῶν παιδίων τε καὶ τῶν ἄλλων τεκμαίρεσθαι τοῖσι ξύμπασι σημείοισιν, ὡς ἐφ᾽ ἑκάστοισιν ἕκαστα διαγέγραπται. ταῦτα δὲ λέγω περὶ τῶν ὀξέων νοσημάτων καὶ ὅσα ἐκ τούτων γίνεται.

25. Χρὴ δὲ τὸν μέλλοντα ὀρθῶς προγινώσκειν τούς τε περιεσομένους καὶ τοὺς ἀποθανουμένους οἷσί τε ἂν μέλλῃ πλείονας ἡμέρας παραμένειν τὸ νόσημα καὶ οἷσιν ἂν ἐλάσσους, τὰ σημεῖα ἐκμανθάνοντα πάντα δύνασθαι κρίνειν λογιζόμενον[73] τὰς δυνάμιας αὐτῶν πρὸς ἀλλήλας, ὥσπερ διαγέγραπται περί τε τῶν ἄλλων καὶ τῶν οὔρων καὶ τῶν πτυέλων, ὅταν ὁμοῦ πύον τε ἀναβήσσῃ καὶ χολήν. χρὴ δὲ καὶ τὰς φορὰς τῶν νοσημάτων τῶν ἀεὶ ἐπιδημεύντων ταχέως ἐνθυμέεσθαι τῆς τε ὥρης τὴν κατάστασιν. εὖ μέντοι χρὴ εἰδέναι περὶ τῶν τεκμηρίων καὶ τῶν ἄλλων σημείων καὶ μὴ λανθάνειν,[74] ὅτι ἐν παντὶ ἔτει καὶ πάσῃ ὥρῃ[75] τά
190 τε κακὰ | κακόν τι σημαίνει καὶ τὰ χρηστὰ ἀγαθόν, ἐπεὶ καὶ ἐν Λιβύῃ καὶ ἐν Δήλῳ καὶ ἐν Σκυθίῃ φαίνεται τὰ προγεγραμμένα ἀληθεύοντα σημεῖα. εὖ οὖν χρὴ

panic-stricken and cry, and if their color changes and becomes green, livid, or red. This is most likely to happen in the youngest children, those up to seven years. Older children and adults with fevers are not seized by spasms, unless one of the most serious and unfavorable signs is present as well, as happens in phrenitis. Decide which of children and of the rest will survive and which will perish on the basis of all the signs as they have been described in each of the cases. This is what I have to say about acute diseases, and what follows from them.

25. Anyone who intends to know correctly in advance which patients will survive and which will perish, and in which patients a disease will last for more days and in which for less, must be able, after learning all the signs, to make his judgment by calculating their significance in relation to one another, as has been explained for other signs, and in particular for urines and for sputa when a patient is simultaneously coughing up pus and bile. It is also necessary immediately to take into account the courses of the endemic diseases and the constitution of the particular season. Indeed, a good understanding of both the certain signs and the others is required, and it can never be forgotten that in every year and season bad signs indicate evil and favorable signs indicate good, since in all of Libya, Delos, and Scythia the signs that have been described appear with their true significance. It must be

---

73 *λογ.* MV: *ἐκλογ.* C′

74 *καὶ μὴ λανθάνειν* C′M: om. V. Both C′ and V also add these words just above, after *ἐνθυμέεσθαι.*

75 *ὥρῃ* MV: *χώρῃ* C′

εἰδέναι, ὅτι ἐν τοῖσιν αὐτοῖσι χωρίοισιν οὐδὲν δεινὸν τὸ μὴ οὐ[76] *τὰ πολλαπλάσια ἐπιτυγχάνειν, ἢν ἐκμαθών τις αὐτὰ κρίνειν τε καὶ λογίζεσθαι*[77] *ὀρθῶς ἐπίστηται. ποθέειν δὲ χρὴ οὐδενὸς νοσήματος ὄνομα, ὅ τι μὴ τυγχάνει ἐνθάδε γεγραμμένον· πάντα γάρ ὁκόσα ἐν τοῖσι χρόνοισι τοῖσι προειρημένοισι κρίνεται, γνώσῃ τοῖσιν αὐτοῖσι σημείοισιν.*

[76] *τὸ μὴ οὐ* M: *τὸ μὴ οὐχὶ* C′: om. V
[77] *λογ.* M: *ἐκλογ.* C′V

realized clearly that in the same regions it is no wonder that someone will be correct in the great majority of cases if he has, after learning the signs, understood how to judge and assess their meaning correctly. Do not regret the absence of any disease's name, which happens not to be mentioned here, for all diseases that reach their crisis in the times given you will be able to recognize by the same signs.

REGIMEN IN ACUTE DISEASES

INTRODUCTION

Regimen in Acute Diseases/Barley Gruel/Against the Cnidian Opinions—all three titles were current in antiquity[1]—consists of two parts: a main treatise of sixty-eight chapters appearing here; an *Appendix*, or "spurious addition" (νόθα), of seventy-two chapters, appearing in Loeb *Hippocrates* volume 6.[2]

Erotian records the title *Barley Gruel* in his census of Hippocratic works, includes seventeen terms drawn from the main treatise in his *Glossary*, and refers three times elsewhere (Ξ3, O21, and P1) to the text by the same title.[3] Galen, who knows this title, prefers *Regimen in Acute Diseases*; he includes about a dozen glosses from the main treatise in his *Explanation of Difficult Words in Hippocrates*, referring to it in two entries (κ33 and μ26) as *Regimen in Acute Diseases*,[4] and cites it many times in his writings.[5] Furthermore, he composed a running commentary in four books (= Gal) to the work as a whole including the *Ap-*

[1] See *Testimonien* vol. I, p. 3f.; vol. II,1, p. 1f.; vol. II,2, p. 3; vol. III, p. 3f.

[2] LCL 473, pp. 257–327 = repr. vol. 6, pp. 223–87.

[3] Cf. Nachmanson, pp. 393–400.

[4] See the passages listed under *De dieta acut.* by Perilli in his *Index locorum Hippocraticorum*, p. 410.

[5] See *Testimonien* vol. II,1, pp. 1–20; vol. II,2, pp. 3–24.

pendix, as well as discussing its place in Hippocrates' medical thought in a monograph, *An Account of Regimen in Acute Diseases in Accordance with the Theories of Hippocrates*: the former of these is extant in the original Greek,[6] while the latter is lost in Greek but preserved in an Arabic translation by Ḥunain ibn Isḥāq.[7]

The wide diffusion of the work in antiquity is shown by its many testimonies in Greek and Latin writers.[8] There is an anonymous ninth-century Arabic translation,[9] which has been edited and translated into English by M. C. Lyons as *Kitāb Tadbīr al-amrāḍ al-ḥādda li-Buqrāṭ. Hippocrates: Regimen in Acute Diseases* (Cambridge, 1966). This Arabic translation was subsequently translated twice into Latin, once either by Constantinus Africanus in the eleventh century or by Gerard of Cremona in the twelfth century, and a second time, possibly by Niccolò da Reggio, in the fourteenth century: together, about two hundred manuscripts of these Latin translations dating between the twelfth and the fifteenth centuries are extant, many of them forming part of the *Articella* collection, to which this treatise was added in the thirteenth century.[10]

[6] G. Helmreich, *In Hippocratis De victu acutorum comm. IV*, CMG V 9,1 (Leipzig and Berlin, 1914), pp. xxvi–xxxviii and 115–366 (= Helmreich). See Manetti/Ros., pp. 1542–545.

[7] M. C. Lyons, *On Regimen in Acute Diseases in Accordance with the Theories of Hippocrates*, CMG Suppl. Orient. II (Berlin, 1969), pp. 12–21 and 74–111. This title is included in Galen's *On My Own Books* (Boudon, p. 162).

[8] See *Testimonien* vol. I, pp. 3–14; vol. III, pp. 3–10.

[9] See Ullmann, p. 29; Sezgin, pp. 33f.

[10] See Kibre, pp. 5–25; Kristeller, p. 70; García-Bal., p. 58.

Regimen in Acute Diseases is structured in chapters as follows:

1–3	Criticism of ancient writers, particularly the authors of *Cnidian Opinions*, for omitting information in their disease accounts, for employing too few medications, for giving no account of regimen, and for classifying specific diseases incorrectly.[11]
4–9	General introduction to regimen in acute diseases, arguing that the present account is necessary because differences among practitioners on the subject have undermined the laity's trust in the medical art.
10–49	Practical instructions for the correct preparation and administration of barley gruel in acute diseases, taking into account the diseases' particular stage of development, as well as patients' signs and symptoms, and their normal regimen.
50–64	Account of the roles wine (50–52), melicrat (53–57), oxymel (58–61), and water (62–63) play in the therapy of acute diseases.
65–68	Benefits of the bath in acute diseases.

[11] These three chapters, which Wilamowitz (p. 16) holds to be alien to the rest of the treatise, have played a large role in the historiography of the Hippocratic Collection for over two millennia. See, e.g., Jouanna *Arch.*, Grensemann *I/II*, Smith *Coans*, Lonie *Cos*, and Thivel.

Into this structure are woven a number of chapters dealing wholly or partly with other topics, for example:

1–9, 26–27, 40–44, 51	Polemical remarks directed against both physicians and laymen for their ignorance and incompetence.[12]
5, 14, 62, 66	Explanations defining and differentiating specific acute diseases.
21–24, 50–53, 62	Accounts of particular therapeutic procedures (fomentations, venesection) and medicines (purgatives, diuretics) valuable in acute diseases.
28–37	Examples of the effects of change of regimen in the healthy that throw light both on disease causation and on the management of regimen in acute diseases.

The text's often associative manner of composition—Galen speaks at one point in his commentary of a "lack of order in the arguments" (ἀταξία τῶν λόγων)[13]—has led many scholars over the ages to hypothesize that the work lacks its final redaction by the author, although the work's overall integrity has not often been called into question.

Besides being printed in all the collected Hippocrates editions and translations, including Adams, Kühlewein, and Chadwick, *Regimen in Acute Diseases* was the subject of a number of special studies, extending from the

[12] Cf. Ducatillon, pp. 285–340.

[13] See Helmreich (n. 6 above), pp. 216f., where Galen is referring to ch. 49 of *Regimen in Acute Diseases*.

sixteenth to the mid-nineteenth century (Littré, vol. 2, pp. 219–22). Important later scholarship on the work includes:

Blum, R. "La composizione dello scritto ippocrateo περὶ διαίτης ὀξέων." *Rendiconti Lincei* 12 (1936): 39–84.

Diller, H. Review of Blum in *Gnomon* 14 (1938): 297–305 (repr. in Diller *KSAM*, pp. 170–77).

Regenbogen, O. "Probleme um die hippokratische Schrift *De victu acutorum*." In *Studies presented to D. M. Robinson*, vol. 2, pp. 624–34. St. Louis, 1953 (repr. in Otto Regenbogen, *Kleine Schriften* [Munich, 1961], pp. 195–205. (= Regenbogen)

Lonie, I. "The Hippocratic Treatise περὶ διαίτης ὀξέων." *Sudhoffs Archiv* 49 (1965): 50–79.

Joly, R. Hippocrate. *Du régime des maladies aiguës*. . . . Budé VI (2). Paris, 1972. (= Joly)

The present edition is based on Jones' Loeb *Hippocrates* and the other studies, in particular Joly's edition. The independent Greek manuscripts have been consulted occasionally on microfilm.

ΠΕΡΙ ΔΙΑΙΤΗΣ ΟΞΕΩΝ[1]

II 224 Littré

1. Οἱ ξυγγράψαντες τὰς Κνιδίας καλεομένας γνώμας ὁκοῖα μὲν πάσχουσιν οἱ κάμνοντες ἐν ἑκάστοισι τῶν νοσημάτων ὀρθῶς ἔγραψαν καὶ ὁκοίως ἔνια ἀπέβαινεν—καὶ ἄχρι μὲν τούτων, καὶ ὃ[2] μὴ ἰητρὸς δύναιτ᾽ ἂν ὀρθῶς ξυγγράψαι, εἰ εὖ παρὰ τῶν καμνόντων ἑκάστου πύθοιτο, ὁκοῖα πάσχουσιν—ὁκόσα δὲ προσκαταμαθεῖν[3] δεῖ τὸν ἰητρὸν μὴ λέγοντος τοῦ κάμνοντος, τούτων πολλὰ παρεῖται, ἄλλ᾽ ἐν ἄλλοισιν καὶ ἐπίκαιρα ἔνια ἐόντα ἐς τέκμαρσιν.

2. Ὁκόταν δὲ ἐς τέκμαρσιν λέγηται, ὡς χρὴ ἕκαστα ἰητρεύειν, ἐν τούτοισι πολλὰ ἑτεροίως | γινώσκω ἢ ὡς ἐκεῖνοι ἐπεξήεσαν. καὶ οὐ μοῦνον διὰ τοῦτο οὐκ ἐπαινέω, ἀλλ᾽ ὅτι καὶ ὀλίγοισι τὸν ἀριθμὸν τοῖσιν ἄκεσιν

226

[1] Περὶ διαίτης ὀξέων. Οἱ δὲ Περὶ πτισάνης. Οἱ δὲ Πρὸς τὰς Κνιδίας γνώμας M: Περὶ πτισάνης A: Πρὸς τὰς Κνιδίας γνώμας ἢ Περὶ πτισάνης V

[2] ὃ Wilamowitz (in Kühlewein): ἦν A: om. MV

[3] προσκ. AV: προκ. M

REGIMEN IN ACUTE DISEASES[1]

1. The authors of the so-called *Cnidian Opinions* wrote correctly about what patients suffer in each of the diseases, and how some of the diseases end—although as far as that goes, even a person who was not a physician could give a correct account like this, if he inquired carefully from each patient what they were suffering—but many of the things a physician must discover beyond this, without the patient telling him, are left out, different things in different cases, some of which are important as clinical evidence.

2. When evidence is being weighed to determine how particular treatments should be applied, in many cases my own position diverges strongly from what is described by these authors. And it is not solely for this reason that I cannot give them my approval, but also because they make use of only a very small number of treatments: indeed,

[1] The titles given by the three independent Greek manuscripts are as follows:

M: *Regimen in Acute Diseases*. Others (sc. say) *Barley Gruel*. Others, *Against the Cnidian Opinions*

A: *Barley Gruel*

V: *Against the Cnidian Opinions* or *Barley Gruel*.

ἐχρέοντο· τὰ γὰρ πλεῖστα αὐτοῖσιν εἴρηται,[4] πλὴν τῶν ὀξειῶν νούσων, φάρμακα ἐλατήρια διδόναι καὶ ὀρὸν καὶ γάλα τὴν ὥρην πιπίσκειν.

3. Εἰ μὲν οὖν ταῦτα ἀγαθὰ ἦν καὶ ἁρμόζοντα τοῖσι νοσήμασιν, ἐφ᾿ οἷσι παρῄνεον διδόναι, πολὺ[5] ἂν ἀξιώτερα ἦν ἐπαίνου, ὅτι ὀλίγα ἐόντα αὐτάρκεά ἐστιν· νῦν δὲ οὐχ οὕτως ἔχει. οἱ μέντοι ὕστερον ἐπιδιασκευάσαντες ἰητρικώτερον δή τι ἐπῆλθον περὶ τῶν προσοιστέων ἑκάστοισιν. ἀτὰρ οὐδὲ περὶ διαίτης οἱ ἀρχαῖοι ξυνέγραψαν οὐδὲν ἄξιον λόγου· καί τοι μέγα τοῦτο παρῆκαν. τὰς μέντοι πολυτροπίας τὰς ἐν ἑκάστῃ τῶν νούσων καὶ τὴν πολυσχιδίην αὐτῶν οὐκ
228 ἠγνόεον | ἔνιοι· τοὺς δ᾿ ἀριθμοὺς ἑκάστου τῶν νοσημάτων σάφα ἐθέλοντες φράζειν οὐκ ὀρθῶς ἔγραψαν· μὴ γὰρ[6] οὐκ εὐαρίθμητον ᾖ, εἰ τούτῳ τις σημαίνεται[7] τὴν τῶν καμνόντων νοῦσον, τῷ τὸ[8] ἕτερον τοῦ ἑτέρου διαφέρειν τι, μὴ τωὐτὸ δὲ νόσημα δοκέειν εἶναι, ἢν μὴ τωὐτὸ ὄνομα ἔχῃ. |

230 4. (2 L.) Ἐμοὶ δὲ ἁνδάνει μὲν ἐν πάσῃ τῇ τέχνῃ προσέχειν τὸν νόον· καὶ γὰρ ὁκόσα ἔργα καλῶς ἔχει ἢ ὀρθῶς, καλῶς ἕκαστα χρὴ ποιέειν καὶ ὀρθῶς, καὶ ὁπόσα ταχέως, ταχέως, καὶ ὁκόσα καθαρίως, καθα-
232 ρίως, καὶ ὁκόσα ἀνωδύνως, διαχειρίζεσθαι | ὡς ἀνωδυνώτατα ποιέειν, καὶ τἆλλα πάντα τοιουτότροπα διαφερόντως τῶν πέλας ἐπὶ τὸ βέλτιον ποιέειν χρή.

[4] εἴρηται A: εὑρέαται M: εὕρηται V
[5] πολὺ AV: ἔτι M
[6] καὶ add. MV

most of what they prescribe—except in acute diseases—is limited to giving purgative medications, and to having patients drink whey and milk at the appropriate times.

3. Now, if these measures were effective and suited to the conditions for which they were prescribed, they would merit much more praise, since, although small in number, they would be adequate for their purpose—but in reality they are not. The writers who later made a revision did in fact come closer to a proper medical understanding of what should be prescribed in each situation. But about regimen, the ancient writers wrote nothing of account, which is a serious omission. Admittedly, some of these authors were not ignorant of the many varieties of each of the diseases and their divisions, but although they intended to set out clearly how many varieties there are of each disease, their account is incorrect: for it would certainly not be easy to calculate (sc. this number) if someone defined the diseases in patients by the fact that one differed from another in some one particular, and if he did not hold a disease to be the same one unless it had the same name.

4. The course I prefer to follow is to pay attention to the entire art (sc. of medicine). Any measures that need to be applied properly and correctly you must execute properly and correctly, if quickly, then quickly, if cleanly, then cleanly, if without pain, you must perform them with the least pain, and everything else like this you must carry out in a superior manner, setting yourself apart from other practitioners.

---

[7] *σημαίνεται* Kühlewein: *σημανεῖται* MV: *σημαίνηται* A
[8] *τῷ τὸ* Gomperz (in Kühlewein): *τῷ* Gal: *τὸ* codd.

5. Μάλιστα δ' ἂν ἐπαινέσαιμι ἰητρόν, ὅστις ἐν τοῖσιν ὀξέσι νοσήμασιν, ἃ τοὺς πλείστους τῶν ἀνθρώπων κτείνει, ἐν τούτοισι διαφέρων τι τῶν ἄλλων εἴη ἐπὶ τὸ βέλτιον. ἔστι δὲ ταῦτα ὀξέα, ὁκοῖα ὠνόμασαν οἱ ἀρχαῖοι πλευρῖτιν καὶ περιπλευμονίην καὶ φρενῖτιν[9] καὶ καῦσον, καὶ τἆλλα ὁκόσα τούτων ἐχόμενα, ὧν οἱ πυρετοὶ τὸ ἐπίπαν ξυνεχέες. ὅταν γὰρ μὴ λοιμώδεος νούσου τρόπος τις κοινὸς ἐπιδημήσῃ, ἀλλὰ σποράδεες ἔωσιν αἱ νοῦσοι, καὶ παραπλάσιοι |
234 ὑπὸ τούτων τῶν νοσημάτων ἀποθνῄσκουσι <ἢ>[10] πλείους ἢ ὑπὸ τῶν ἄλλων τῶν ξυμπάντων.

6. Οἱ μὲν οὖν ἰδιῶται οὐ κάρτα γινώσκουσιν τοὺς
236 ἐς ταῦτα διαφέροντας τῶν πέλας ἑτεροίων τε | μᾶλλον ἐπαινέται ἰημάτων καὶ ψέκται εἰσίν· ἐπεί τοι μέγα σημεῖον τόδε, ὅτι οἱ δημόται ἀξυνετώτατοι αὐτοὶ ἑωυτῶν περὶ τούτων τῶν νοσημάτων εἰσίν, ὡς μελετητέα ἐστί·[11] οἱ γὰρ μὴ ἰητροὶ ἰητροὶ δοκέουσιν εἶναι μάλι-
238 στα διὰ | ταύτας τὰς νούσους· ῥηΐδιον γὰρ τὰ ὀνόματα ἐκμαθεῖν, ὁκοῖα νενόμισται προσφέρεσθαι πρὸς τοὺς τὰ τοιάδε κάμνοντας· ἢν γὰρ ὀνομάσῃ τις πτισάνης τε χυλὸν καὶ οἶνον τοῖον ἢ τοῖον καὶ μελίκρητον, πάντα τοῖσιν ἰδιώτῃσι[12] δοκέουσιν οἱ ἰητροὶ τὰ αὐτὰ λέγειν, οἵ τε βελτίους καὶ οἱ χείρους. τὰ δὲ οὐχ οὕτως ἔχει, ἀλλ' ἐν τούτοισι δὴ καὶ πάνυ μέγα[13] διαφέρουσιν ἕτεροι ἑτέρων.

[9] καὶ λήθαργοι add. M [10] Conj. Regenbogen (p. 627, n. 2) [11] ἐστί Kühlewein: εἶναι codd.
[12] ἰδιώτῃσι A: δημότῃσι MV

5. I would value most highly the physician who in acute diseases—which kill the most people—distinguished himself in some way as better than others. These acute diseases are the ones the ancient writers called pleurisy, pneumonia, phrenitis and ardent fever,[2] besides other conditions like these whose fevers are generally continuous. For when no kind of shared pestilential disease is present in a population, but the diseases are sporadic, about the same number of people die from acute diseases, or even more, than from all the rest together.

6. Now laymen do not distinguish very well between practitioners whose competence in such matters excels, and their fellows, and tend rather to give praise or to find fault on the basis of exotic remedies. In fact, there is much evidence to suggest that common people on their own are quite incapable of understanding this sort of disease, as far as the management is concerned, for it is from (sc. their experience with) such diseases that they think false physicians are real physicians. After all, it is easy to pick up the names of treatments that are usually given in such cases, so that when someone names barley water, and such and such a wine, and melicrat, laymen think that both competent physicians and incompetent physicians say the same things. But this is not at all so, since practitioners differ among themselves very greatly in these matters.

[2] This same list is also found at *Affections* 6. For alternatives, see *Airs Waters Places* 3 (pleurisies, pneumonias, ardent fevers), *Prognostic* 4 (acute fevers, pneumonias, phrenitises, headaches), *Coan Prenotions* 487 (pneumonias, pleurisies).

---

13 μέγα M: μεγάλα V: om. A

7. (3 L.) Δοκέει δέ μοι ἄξια γραφῆς εἶναι,[14] ὁκόσα τε ἀκαταμάθητά ἐστιν τοῖς ἰητροῖσιν ἐπίκαιρα ἐόντα εἰδέναι καὶ ὁκόσα μεγάλας ὠφελείας φέρει ἢ μεγάλας βλάβας. ἀκαταμάθητα μεν[15] οὖν καὶ τάδ᾽ ἐστίν, διὰ τί ἄρα ἐν τῇσιν ὀξείῃσι νούσοισιν οἱ μὲν τῶν ἰητρῶν πάντα τὸν αἰῶνα διατελέουσι πτισάνας διδόντες ἀδι-
240 ηθήτους καὶ νομίζουσιν | ὀρθῶς ἰητρεύειν, οἱ δέ τινες περὶ παντὸς ποιέονται, ὅπως κριθὴν μηδεμίαν καταπίῃ ὁ κάμνων—μεγάλην γὰρ βλάβην ἡγεῦνται εἶναι—ἀλλὰ δι᾽ ὀθονίου τὸν χυλὸν διηθέοντες διδόασιν· οἱ δ᾽ αὖ τινες αὐτῶν οὔτ᾽ ἂν πτισάνην παχεῖαν δοῖεν οὔτε χυλόν· οἱ μὲν μέχρι ἂν ἑβδομαῖος γένηται, οἱ δὲ καὶ διὰ τέλεος ἄχρι ἂν κριθῇ ἡ νοῦσος.

8. Μάλα μὲν οὖν οὐδὲ προβάλλεσθαι τὰ τοιαῦτα ζητήματα εἰθισμένοι εἰσὶν οἱ ἰητροί. ἴσως δὲ οὐδὲ προβαλλόμενα γινώσκεται·[16] καίτοι διαβολήν γε ἔχει ὅλη ἡ τέχνη πρὸς τῶν δημοτέων μεγάλην, ὡς μὴ δο-
242 κέειν ὅλως ἰητρικὴν εἶναι· ὥστ᾽ εἰ ἔν γε τοῖσιν | ὀξυτάτοισι τῶν νοσημάτων τοσόνδε διοίσουσιν ἀλλήλων οἱ χειρώνακτες, ὥστε ἃ ὁ ἕτερος προσφέρει ἡγεύμενος ἄριστα εἶναι, ταῦτα νομίζειν[17] τὸν ἕτερον κακὰ εἶναι, σχεδὸν ἂν κατά γε τῶν τοιούτων τὴν τέχνην φαῖεν ὡμοιῶσθαι μαντικῇ, ὅτι καὶ οἱ μάντιες τὸν αὐτὸν ὄρνιθα, εἰ μὲν ἀριστερὸς εἴη, ἀγαθὸν νομίζουσιν εἶναι, εἰ δὲ δεξιός, κακόν—καὶ ἐν ἱεροσκοπίῃ[18] δὲ τοιάδε,

[14] Add. μάλιστα MV [15] μὲν MV: om. A
[16] γινώσκεται A: εὑρίσκεται MV

7. I also believe that it is worthwhile to write about the subjects on which physicians disagree, but which are vital to understand, and which result in great benefit or great harm. It is a point of disagreement, for example, that in acute diseases some physicians continue through the whole course to administer unstrained barley gruel, believing that they are giving the proper treatment, while other physicians go to great lengths to prevent patients from swallowing the least barley—fearing that it would do serious harm—by giving barley water that has been filtered through a linen cloth. Then again, there are others who give neither unstrained barley gruel nor the filtered juice; and others who give nothing at all until the seventh day, and yet others who give nothing before the disease has reached its crisis.

8. Now physicians are not even in the habit of bringing up questions like these, and if they did bring them up, I doubt that they would be resolved. Yet this reticence gives the whole art a very bad name among the laity, leading them to have doubts whether it exists at all. Thus, if in the acutest diseases practitioners disagree so much among themselves, with what one of them prescribes in the belief that it is the best treatment being regarded by another as a bad treatment, is it any wonder that to the laity medicine looks like fortune-telling? For some fortune-tellers hold that if a particular bird appears from the left side it is an omen of good fortune, but if from the right side a bad omen—the situation is the same with augury from animal

[17] Add. δὴ Gal: add. ἤδη MV
[18] ἱερο- AV: ἀερο- M

244 *ἄλλα ἐπ᾽ ἄλλοις—ἔνιοι δὲ τῶν μαντίων* | *τὰ ἐναντία τούτων.*

9. *Φημὶ δὲ πάγκαλον εἶναι τοῦτο τὸ σκέμμα καὶ ἠδελφισμένον τοῖσι πλείστοισι τῶν ἐν τῇ τέχνῃ καὶ ἐπικαιροτάτοισι· καὶ γὰρ τοῖσι νοσέουσι πᾶσιν ἐς ὑγιείην μέγα τι δύναται καὶ τοῖσιν ὑγιαίνουσιν ἐς ἀσφάλειαν καὶ τοῖσιν ἀσκέουσιν ἐς εὐεξίην καὶ ἐς ὅ τι ἕκαστος ἐθέλοι.*

10 (4 L.) *Πτισάνη μὲν οὖν μοι*[19] *δοκέει ὀρθῶς προκεκρίσθαι τῶν σιτηρῶν γευμάτων ἐν τούτοισι τοῖσι νοσήμασι, καὶ ἐπαινέω τοὺς προκρίναντας. τὸ γὰρ γλίσχρασμα αὐτῆς λεῖον καὶ ξυνεχὲς καὶ προσηνές* 246 *ἐστι καὶ ὀλισθηρὸν καὶ πλαδαρὸν μετρίως καὶ* | *ἄδιψον καὶ εὐέκκριτον, εἴ τι καὶ τούτου προσδέοι, καὶ οὔτε στύψιν ἔχον οὔτε ἄραδον κακὸν οὔτε ἀνοιδίσκεται ἐν τῇ κοιλίῃ· ἀνῴδηκε γὰρ ἐν τῇ ἑψήσει, ὁκόσον πλεῖστον ἐπεφύκει διογκοῦσθαι.*

11. *Ὁκόσοι μὲν οὖν πτισάνῃσι χρέονται ἐν τούτοισι τοῖσι νοσήμασιν, οὐδεμιῇ ἡμέρῃ κενεαγγητέον, ὡς ἔπος εἰρῆσθαι, ἀλλὰ χρηστέον καὶ οὐ διαλειπτέον, ἢν μή τι δέῃ ἢ διὰ φαρμακείην ἢ κλύσιν διαλείπειν. καὶ τοῖσι μέν γε εἰθισμένοισι δὶς σιτέεσθαι τῆς ἡμέρης δὶς δοτέον· τοῖσι δὲ μονοσιτέειν εἰθισμένοισι ἅπαξ δοτέον τὴν πρώτην· ἐκ προσαγωγῆς* [*ἢν*][20] *δ᾽ ἐνδέχεται* 248 *καὶ τούτοισι δὶς διδόναι, ἢν* | *δοκέῃ προσδεῖν. πλῆθος δὲ ἀρκέει κατ᾽ ἀρχὰς διδόναι μὴ πολὺ μηδὲ ὑπέρπαχυ, ἀλλ᾽ ὁκόσον εἵνεκα τοῦ ἔθεος ἐσιέναι τι καὶ κενεαγγίην μὴ γενέσθαι πολλήν.*

entrails, too, *mutatis mutandis*—whereas other fortune-tellers believe just the opposite.

9. But I insist that this debate is completely fitting, and concerns many of the most critical questions the art is faced with. It is very significant for restoring health in all diseases, for providing safety in persons enjoying good health, for achieving top condition in athletes, and in fact to promote what each particular person might wish for.

10. Now barley gruel has, in my opinion, been rightly preferred among the cereal preparations for acute diseases, and I share this preference. For its gluten is smooth, consistent, soothing, lubricant, and moderately moistening; it relieves thirst, and it is easily excreted, if this action is also required. It is neither astringent nor irritating, nor does it swell up inside the cavity, since it has already swollen to its maximum natural volume when it was boiled.

11. Patients who employ barley gruel during acute diseases should in general avoid fasting on any day, but continue the use of gruel without interruption, unless some break is needed because of a purgation or an enema. To those who habitually eat twice a day, it should be given twice a day, to those who eat once a day, once on the first day, but, if it seems advisable, it may be gradually increased to twice a day in such patients, too. As for the amount, at the beginning it is sufficient to give neither very much nor very thick gruel, but just as much as the person's habit indicates, and such as to avoid any great hunger.

---

[19] μοι MV: om. A
[20] ἦν del. Reinhold

12. Περὶ δὲ τῆς ἐπιδόσιος ἐς πλῆθος τοῦ ῥυφήματος, ἢν μὲν ξηρότερον ᾖ τὸ νόσημα ἢ ὡς ἄν τις
250 βούληται,[21] οὐ χρὴ ἐπὶ πλέον διδόναι, | ἀλλὰ προπίνειν πρὸ τοῦ ῥυφήματος ἢ μελίκρητον ἢ οἶνον, ὁκότερον ἂν ἁρμόζῃ· τὸ δ' ἁρμόζον ἐφ' ἑκάστοισι τῶν τρόπων εἰρήσεται. ἢν δὲ ὑγραίνηται τὸ στόμα καὶ τὰ ἀπὸ τοῦ πλεύμονος ἴῃ ὁκοῖα δεῖ, ἐπιδιδόναι χρὴ ἐς πλῆθος τοῦ ῥυφήματος, ὡς ἐν κεφαλαίῳ εἰρῆσθαι· τὰ μὲν γὰρ θᾶσσον καὶ μᾶλλον πλαδῶντα ταχυτῆτα σημαίνει κρίσιος, τὰ δὲ βραδύτερον πλαδῶντα καὶ ἧσσον βραδυτέρην σημαίνει τὴν κρίσιν. καὶ ταῦτα αὐτὰ μὲν καθ' ἑωυτὰ τοιάδε τὸ ἐπίπαν ἐστί.

13. Πολλὰ δὲ καὶ ἄλλα ἐπίκαιρα παρεῖται, οἷσι προσημαίνεσθαι δεῖ, ἃ εἰρήσεται ὕστερον. καὶ ὁκόσῳ ἂν πλείων ἡ κάθαρσις γίνηται, τοσῷδε χρὴ πλεῖον διδόναι ἄχρι κρίσιος· μάλιστα δὲ κρίσιος ὑπερβολῆς δύο ἡμερέων, οἷσί γε ἢ πεμπταίοισιν ἢ ἑβδομαίοισιν
252 ἢ ἐναταίοισι δοκέει κρίνεσθαι, ὡς καὶ τὸ ἄρτιον | καὶ τὸ περισσὸν προμηθὲς ᾖ· μετὰ δὲ τοῦτο τῷ μὲν ῥυφήματι τὸ πρωῒ χρηστέον, ἐς ὀψὲ δὲ ἐς σιτία μεταβάλλειν.

14. Ξυμφέρει δὲ τὰ τοιάδε ὡς ἐπὶ τὸ πολὺ τοῖς
254 οὔλῃσι πτισάνῃσιν αὐτίκα | χρεωμένοισιν. αἵ τε γὰρ ὀδύναι ἐν τοῖσι πλευριτικοῖσιν αὐτίκα παύονται αὐτόματοι, ὅταν ἄρξωνται πτύειν τι ἄξιον λόγου καὶ ἐκκαθαίρεσθαι, αἵ τε καθάρσιες πολλῷ τελεώτεραί εἰσι, καὶ ἔμπυοι ἧσσον γίνονται, ἢ εἰ ἀλλοίως τις διαιτῴη,

12. As for increasing the amount of gruel given, if the disease is drier than desired, the dose of gruel should not be increased, but rather, before the gruel have the patient take a drink of either melicrat or wine, whichever seems more appropriate—what is appropriate in each case is explained below. But if the patient's mouth is moist, and his sputa are passing as they should, in principle you should increase the amount of gruel. For moistness arriving sooner and in a greater volume indicates an earlier crisis, whereas moistness coming later and in a smaller amount indicates a later crisis. In essence, the matter is generally so.

13. Many other critical subjects are omitted here, which play a role in prognosis, but they will be handled later. The greater the amount (sc. of sputum) evacuated, the more (sc. barley gruel) should be given, up to the crisis; then you should continue with gruel for two days after the crisis, especially in patients whose crisis is revealed on the fifth, seventh or ninth day, in order that both the even and the odd day will be covered. After that you should employ gruel early in the day, but later change to foods.

14. These (sc. rules) are generally beneficial for patients who begin to employ unstrained barley gruel immediately. In cases of pleurisy, the pains stop spontaneously very soon, when the patients begin to expectorate a significant amount of sputum and to be cleaned. Their cleaning is much completer, and they develop fewer suppurations than if any other regimen is followed; their

---

21 *βούληται* A: *οἴοιτο* MV

καὶ αἱ κρίσιες ἁπλούστεραι καὶ εὐκριτώτεραι καὶ ἧσσον ὑποστροφώδεες.

15. (5 L.) Τὰς *δὲ πτισάνας χρὴ ἐκ κριθέων ὡς βελτίστων εἶναι καὶ κάλλιστα ἑψῆσθαι, καὶ ἄλλως ἢν μὴ τῷ χυλῷ μούνῳ μέλλῃς χρέεσθαι· μετὰ γὰρ*
256 *τῆς ἄλλης ἀρετῆς τῆς πτισάνης τὸ ὀλισθηρὸν* | *τὴν κριθὴν καταπινομένην ποιέει μὴ βλάπτειν· οὐδαμῇ γὰρ προσίσχει οὐδὲ μένει κατὰ τὴν τοῦ θώρηκος ἴξιν· ὀλισθηροτάτη δὲ καὶ ἀδιψοτάτη καὶ εὐπεπτοτάτη καὶ ἀσθενεστάτη ἐστὶν ἡ κάλλιστα ἑφθή· ὧν πάντων δεῖ.*

16. Ἢν *οὖν μὴ προστιμωρήσῃ τις ὁκόσων δεῖται αὐτάρκης εἶναι ὁ τρόπος τῆς τοιαύτης πτισανορρυφίης, πολλαχῇ βεβλάψεται. ὁκόσοισι γὰρ σῖτος αὐτίκα ἐγκατακέκλεισται, εἰ μή τις ὑποκενώσας δοίη τὸ ῥύφημα, τὴν ὀδύνην ἐνεοῦσαν προσπαροξύνειεν*
258 *ἂν* | *καὶ μὴ ἐνεοῦσαν ἐμποιήσειεν ἄν,*[22] *καὶ πνεῦμα πυκνότερον γένοιτ᾽ ἄν· κακὸν δὲ τοῦτο· ξηραντικόν τε γὰρ πλεύμονος καὶ κοπῶδες ὑποχονδρίων καὶ ἤτρου καὶ φρενῶν· τοῦτο δέ, ἢν ἔτι τοῦ πλευροῦ τῆς ὀδύνης ξυνεχέος ἐούσης καὶ πρὸς τὰ θερμάσματα μὴ χαλώσῃς καὶ τοῦ πτυάλου μὴ ἀνιόντος, ἀλλὰ καταγλι-*
260 *σχραινομένου ἀσαπέως, ἢν μὴ λύσῃ τις* | *τὴν ὀδύνην ἢ κοιλίην μαλθάξας ἢ φλέβα ταμών, ὁκότερον ἂν τούτων σημήνῃ,*[23] *τὰς δὲ πτισάνας ἢν οὕτως ἔχωσι διδῷ, ταχέες οἱ θάνατοι τῶν τοιούτων γίνονται.*

[22] *ἐμποιήσειεν ἂν* A: *εὐθὺς ποιήσειεν* MV
[23] *σημήνῃ* A: *ξυμφέρει* M: *ξυμφέρῃ* V

crises are completer, more decisive, and less subject to relapse.

15. Gruels must be made from the best barley, and very thoroughly boiled, especially if you are intending to use not just the barley water. For along with the other benefits of gruel, its slipperiness prevents the barley which is being ingested from doing any harm, since it can never catch or stick on its way through the chest. The most perfectly boiled gruel lubricates best, produces the least thirst, and is the most digestible and mild—all properties that are required.

16. Now if someone fails to pay attention to the conditions which the drinking of this kind of barley gruel requires in order to be beneficial, many kinds of injury will result. Patients in whom food is immediately confined (sc. in the cavity)—if the gruel is administered without first inducing a downward cleaning—will then suffer an exacerbation of any pain that is already present, or, if they have no pain, a new pain will be provoked, and their breathing will become more rapid: this is bad because it causes a drying of the lung as well as pain in the hypochondrium, lower abdomen, and diaphragm. Furthermore, for example, when pains in the side persist and do not cede in response to fomentations, and the sputum is not coughed up, but coagulates in an unconcocted mass, if no one allays the pain by softening the cavity, or incising a vessel—whichever of these measures is indicated—but gives barley gruel to any such patient, death will soon follow.

17. Διὰ ταύτας οὖν τὰς προφάσιας καὶ ἑτέρας τοιαύτας[24] οἱ οὔλῃσι πτισάνῃσι χρεώμενοι ἑβδομαῖοι καὶ ὀλιγημερώτεροι θνῄσκουσιν, οἱ μέν τι καὶ τὴν γνώμην βλαβέντες, οἱ δ' ὑπὸ τῆς ὀρθοπνοίης τε καὶ τοῦ ῥέγχεος ἀποπνιγέντες. μάλα δὲ τοὺς τοιούτους οἱ

262 ἀρχαῖοι βλητοὺς ἐνόμιζον εἶναι διὰ τόδε | οὐχ ἥκιστα, ὅτι ἀποθανόντων αὐτῶν ἡ πλευρὴ πελιδνὴ εὑρίσκεται, ἴκελόν τι πληγῇ. αἴτιον δὲ τούτου ἐστίν, ὅτι πρὶν λυθῆναι τὴν ὀδύνην θνῄσκουσιν· ταχέως γὰρ πνευματίαι γίνονται· ὑπὸ δὲ τοῦ πολλοῦ καὶ πυκνοῦ πνεύματος, ὡς ἤδη εἴρηται, καταγλισχραινόμενον τὸ πτύαλον ἀπέπτως κωλύει τὴν ἐπάνοδον γίνεσθαι, ἀλλὰ τὴν ῥέγξιν ποιέει ἐνισχόμενον ἐν τοῖσι βρογχίοισι τοῦ πνεύμονος. καὶ ὁκόταν ἐς τοῦτο ἔλθῃ, θανατῶδες ἤδη ὡς ἐπὶ τὸ πολύ ἐστι· καὶ γὰρ αὐτὸ τὸ πτύαλον ἐνισχόμενον κωλύει μὲν τὸ πνεῦμα ἔσω φέρεσθαι, ἀναγκάζει δὲ ταχέως ἔξω φέρεσθαι· καὶ οὕτως ἐς τὸ κακὸν ἀλλήλοισι συντιμωρεῖ.[25] τό τε γὰρ πτύαλον ἐνισχόμενον πυκνὸν τὸ πνεῦμα ποιέει, τό τε πνεῦμα πυκνὸν ἐὸν ἐπιγλισχραίνει τὸ πτύαλον καὶ κωλύει

264 ἀπολισθάνειν. καταλαμβάνει | δὲ ταῦτα οὐ μοῦνον ἢν πτισάνῃ ἀκαίρως χρέωνται, ἀλλὰ πολὺ μᾶλλον, ἤν τι ἄλλο φάγωσιν ἢ πίωσι πτισάνης ἀνεπιτηδειότερον.

18. (6 L.) Μάλα μὲν οὖν τὰ πλεῖστα παραπλήσιοί εἰσιν αἱ τιμωρίαι τοῖσί τε οὔλῃσι πτισάνῃσι χρεωμένοισι τοῖσί τε χυλῷ αὐτῷ· τοῖσι δὲ μηδετέρῳ τούτων, ἀλλὰ ποτῷ μοῦνον, ἔστι δὲ ὅπῃ καὶ διαφερόντως τιμωρητέον. χρὴ δὲ τὸ πάμπαν οὕτω ποιέειν.

17. Now for these and similar reasons, people taking unstrained barley gruel tend to die on the seventh day or before, some falling into delirium, and others being suffocated with orthopnea and stertorous breathing. Generally, the ancient writers thought that such patients were stricken, not least because when they died their side was discovered to have a livid mark, as if from a blow. This is because they died before their pain went away, as dyspnea quickly developed with deep and rapid breathing, as noted above. The sputum, being unconcocted, becomes very viscid, which stops its expectoration from proceeding, and the stertorous breathing causes it to be held back in the bronchi and the lung. Once the disease reaches this state, it is usually fatal. The sputum, being held back, prevents the breath from being drawn in, and forces it to be quickly expired. Thus the two difficulties aggravate one another: the sputum, by being held back inside, causes rapid respiration, and the respiration, being rapid, makes the sputum even more viscid, and prevents it from gliding away. These things happen not only if people take barley gruel at the wrong time, but even more if they eat or drink something more unsuitable than barley gruel.

18. Now the measures required when employing unstrained barley gruel are practically the same as for employing just barley water. Patients who take neither of these, but simply a drink, require a somewhat different treatment. In general, you must do as follows.

---

24 Add. ἔτι μᾶλλον M: add. μᾶλλον A
25 συντιμωρεῖ A: τιμωρέουσι MV

19. Ἢν μὲν νεοβρῶτι αὐτῷ ἐόντι καὶ κοιλίης μήπω ὑποκεχωρηκυίης ἄρξηται ὁ πυρετός, ἤν τε ξὺν ὀδύνῃ ἤν τε ἄνευ ὀδύνης, ἐπισχεῖν τὴν δόσιν τοῦ ῥυφήματος, ἔστ᾽ ἂν οἴηται κεχωρηκέναι ἐς τὸ κάτω μέρος τοῦ ἐντέρου τὸ σιτίον. χρέεσθαι δὲ ποτῷ, ἢν μὲν ἄλγημά
266 τι ἔχῃ, ὀξυμέλιτι, χειμῶνος | μὲν θερμῷ, θέρεος δὲ ψυχρῷ· ἢν δὲ πολλὴ δίψα ᾖ, καὶ μελικρήτῳ καὶ ὕδατι. ἔπειτα, ἢν μὲν ἄλγημα ἐνῇ ἢ τῶν ἐπικινδύνων τι ἐμφαίνηται, διδόναι τὸ ῥύφημα μήτε πολὺ μήτε παχύ, μετὰ δὲ τὴν ἑβδόμην ἢ ἐνάτην,[26] ἢν ἰσχύῃ. ἢν δὲ μὴ ὑπεληλύθῃ ὁ παλαιότερος σῖτος νεοβρῶτι ἐόντι, ἢν μὲν ἰσχύῃ τε καὶ ἀκμάζῃ τῇ ἡλικίῃ, κλύσαι, ἢν δὲ ἀσθενέστερος ᾖ, βαλάνῳ προσχρήσασθαι, ἢν μὴ αὐτόματα διεξίῃ καλῶς.

20. Καιρὸν δὲ τῆς δόσιος τοῦ ῥυφήματος τόνδε μάλιστα φυλάσσεσθαι κατ᾽ ἀρχὰς καὶ διὰ παντὸς τοῦ νοσήματος· ὅταν μὲν οἱ πόδες ψυχροὶ ἔωσιν, ἐπισχεῖν χρὴ τοῦ ῥυφήματος τὴν δόσιν, μάλιστα δὲ καὶ τοῦ ποτοῦ ἀπέχεσθαι· ὁκόταν δὲ ἡ θέρμη καταβῇ ἐς
268 τοὺς πόδας, τότε διδόναι· καὶ | νομίζειν μέγα δύνασθαι τὸν καιρὸν τοῦτον ἐν πάσῃσι τῇσι νούσοισιν, οὐχ ἥκιστα δὲ ἐν τῇσιν ὀξείῃσιν, μάλιστα δ᾽ ἐν τῇσι μᾶλλον πυρετώδεσι καὶ ἐπικινδυνοτάτῃσι.[27] χρῆσθαι δὲ πρῶτον μὲν χυλῷ, ἔπειτα δὲ πτισάνῃ, κατὰ τὰ τεκμήρια τὰ προγεγραμμένα ἀκριβῶς θεωρῶν.

21. (7 L.) Ὀδύνην δὲ πλευροῦ, ἤν τε κατ᾽ ἀρχὰς γίνηται ἤν θ᾽ ὕστερον, θερμάσμασι μὲν πρῶτον οὐκ ἀπὸ τρόπου ἐστι χρησάμενον πειρηθῆναι διαλῦσαι.

19. If fever begins when a person has just eaten, and his cavity has not yet had a downward movement—whether pain is present or not—refrain from giving gruel until he believes the food has moved into the lower part of his intestine. If pain is present, employ the following drinks: oxymel, warm in winter and cold in summer; if there is violent thirst, both melicrat and water. Later, if there is some pain, or signs of danger appear, give gruel, neither a great amount nor too thick, and only after the seventh or ninth day, if the patient is strong. But when the earlier food has not descended in a person who has just eaten, if he is strong and at the peak of his age, give him an enema; if he is too weak for that, apply a suppository, if the stools do not pass properly on their own.

20. The opportune moment to give a gruel must be carefully watched for both at the beginning and through the whole disease. When the feet are cold, you should not give any gruel, and especially not a drink; but when warmth moves down into the feet, give one then. Be aware that this movement has great (sc. therapeutic) significance in all diseases, particularly acute ones, and most of all in ones that are more feverish and most dangerous. Employ the juice first, and then the gruel, paying careful attention to the signs I have noted.

21. Pain in the side, whether it occurs at the beginning or later, it is not amiss first to try to allay by applying fomentations. Of these the most powerful one is hot water

---

[26] ἢ ἐνάτην MV: om. A

[27] καὶ ἐπικινδυνοτάτῃσι MV: om. A

θερμασμάτων δὲ κράτιστον μὲν ὕδωρ θερμὸν ἐν ἀσκῷ ἢ ἐν κύστει ἢ ἐν χαλκῷ ἀγγείῳ ἢ ἐν ὀστρακίνῳ. προϋποτιθέναι δὲ | χρὴ μαλθακόν τι πρὸς τὴν πλευρὴν προσηνείης εἵνεκεν. ἀγαθὸν δὲ καὶ σπόγγος μαλθακὸς μέγας ἐξ ὕδατος θερμοῦ ἐκπεπιεσμένος προστίθεσθαι. περιστέγειν δὲ ἱματίῳ[28] *τὴν θάλψιν χρή, πλείω τε γὰρ χρόνον ἀρκέσει καὶ παραμενεῖ. καὶ ἅμα ὡς μὴ ἡ ἀτμὶς πρὸς τὸ πνεῦμα τοῦ κάμνοντος φέρηται, ἢν ἄρα μὴ δοκέῃ καὶ τοῦτο χρήσιμον πρός τι εἶναι· ἔστι γὰρ ὅτε δεῖ πρός τι. ἔτι δὲ καὶ κριθαὶ καὶ ὄροβοι· ἐν ὄξει κεκρημένῳ σμικρῶς ὀξυτέρῳ*[29] *ἢ ὡς ἂν πίοι τις διέντα καὶ ἀναζέσαντα ἐς μαρσίππια καταρράψαντα προστιθέναι. καὶ πίτυρα τὸν αὐτὸν τρόπον. ξηρὴ δὲ πυρίη, ἅλες, κέγχροι πεφρυγμένοι ἐν εἰρινέοισι μαρσιππίοισιν ἐπιτηδειότατοι· καὶ γὰρ κοῦφον καὶ | προσηνὲς ὁ κέγχρος.*

270

272

22. *Λύει δὲ μάλθαξις ἡ τοιῆδε καὶ τὰς πρὸς κληῖδα περαινούσας ἀλγηδόνας· τομὴ μέντοι οὐχ ὁμοίως λύει ὀδύνην, ἢν μὴ πρὸς τὴν κληῖδα περαίνῃ ἡ ὀδύνη· ἢν δὲ μὴ λύηται πρὸς τὰ θερμάσματα ὁ πόνος, οὐ χρὴ πολὺν χρόνον θερμαίνειν· καὶ γὰρ ξηραντικὸν τοῦ πλεύμονος τοῦτ' ἐστι καὶ ἐμπυητικόν· ἀλλ' ἢν μὲν σημαίνῃ ἡ ὀδύνη ἐς κληῖδα ἢ ἐν βραχίονι βάρος ἢ περὶ μαζὸν ἢ ὑπὲρ τῶν φρενῶν, τάμνειν ἀρήγει τὴν ἐν τῷ ἀγκῶνι φλέβα τὴν ἔσω καὶ μὴ ὀκνεῖν συχνὸν ἀφαιρέειν, ἕως ἂν ἐρυθρότερον πολλῷ ῥυῇ ἢ ἀντὶ καθαροῦ τε καὶ ἐρυθροῦ | πελιόν· ἀμφότερα γὰρ γίνεται.*

274

in a skin or bladder, or in a bronze vessel, or in an earthenware pot. You must first interpose some soft object against the patient's side in order to prevent discomfort. It is also good to apply a large, soft sponge taken out of hot water and squeezed out. The fomentation should be covered with a blanket, in order to make it effective for a longer time, and to hold it in place. At the same time this will prevent the vapor from being drawn in with the patient's breath, unless you think that would be beneficial—occasionally it may be necessary for some reason. Also barley and vetches: apply these in vinegar mixed a little stronger than can be imbibed, sieved, boiled and tied up in small pouches. Also bran prepared in the same way. For dry fomentations, salt and toasted millet in small woolen pouches are ideal, since the millet is light and soothing.

22. A softening agent of this kind also relieves pains extending in the direction of the collarbone, whereas (sc. vene-)section does not relieve pain so well, unless the pain reaches all the way to the collarbone itself. If the pain is not relieved by fomentations, do not continue the warming for very long, since that would cause drying of the lung, and produce internal suppuration. If the pain shows signs of moving to the collarbone, or a weight is felt in the forearm, or around a breast, or above the diaphragm, you should incise the inside vessel at the elbow, and not hesitate to continue drawing off (sc. blood) until it flows out much redder than before, or changes from being pure red to being livid—for both these things may happen.

---

28 *ἱματίῳ* MV: *ἄνω* A

29 *ὀξυτέρῳ* Kühlewein: *ὀξύτερον* codd.

23. Ἢν δ' ὑπὸ φρένας ᾖ τὸ ἄλγημα, ἐς δὲ τὴν κληῖδα μὴ σημαίνῃ, μαλθάσσειν χρὴ τὴν κοιλίην ἢ μέλανι ἐλλεβόρῳ ἢ πεπλίῳ, μέλανι μὲν δαῦκος ἢ σέσελι ἢ κύμινον ἢ ἄνησον ἢ ἄλλο τι τῶν εὐωδέων μίσγοντα, πεπλίῳ δὲ ὀπὸν σιλφίου. ἀτὰρ καὶ μισγόμενα ἀλλήλοισιν ὁμοιότροπα ταῦτ' ἐστιν. ἄγει δὲ μέλας μὲν καλλίω καὶ κρισιμώτερα πεπλίου, πέπλιον δὲ μέλανος φυσέων καταρρηκτικώτερόν ἐστιν. ἄμφω δὲ ταῦτα ὀδύνην παύει· παύει δὲ καὶ ἄλλα πολλὰ τῶν ὑπηλάτων· κράτιστα δὲ ταῦτα, ὧν ἐγὼ οἶδα, ἐστίν·
276 ἐπεὶ καὶ τὰ ἐν τοῖσι ῥυφήμασι | διδόμενα ὑπήλατα ἀρήγει, ὁκόσα μὴ ἄγαν ἐστὶν ἀηδέα ἢ διὰ πικρότητα ἢ δι' ἄλλην τινὰ ἀηδίην, ἢ διὰ πλῆθος ἢ διὰ χροιὴν ἢ ὑποψίην τινά.

24. Τῆς μέντοι πτισάνης, ὁκόταν πίῃ τὸ φάρμακον, ἐπιρρυφεῖν αὐτίκα χρὴ διδόναι μηδὲν ἔλασσον ἀξίως λόγου ἢ ὁκόσον εἴθιστο· ἐπεὶ καὶ κατὰ λόγον ἐστὶ μεσηγὺ τῆς καθάρσιος μὴ διδόναι ῥυφεῖν· ὅταν δὲ λήξῃ ἡ κάθαρσις, τότε ἔλασσον ῥυφεέτω ἢ ὁκόσον εἴθιστο. μετὰ δὲ ταῦτα ἀναγέτω αἰεὶ[30] ἐπὶ τὸ πλεῖον, ἢν ἥ τε ὀδύνη πεπαυμένη ᾖ καὶ μηδὲν ἄλλο ἐναντιῶται.

25. Ωὑτὸς δέ μοι λόγος ἐστίν, καὶ ἢν χυλῷ πτισάνης δέῃ χρέεσθαι. φημὶ γὰρ ἄμεινον εἶναι αὐτίκα ἄρξασθαι ῥυφεῖν τὸ ἐπίπαν μᾶλλον ἢ προκενεαγγή-
278 σαντα ἄρξασθαι τοῦ ῥυφήματος ἢ τριταῖον | ἢ τεταρταῖον ἢ πεμπταῖον ἢ ἑκταῖον ἢ ἑβδομαῖον, ἤν γε μὴ

23. If the pain is below the diaphragm, and it gives no sign of moving toward the collarbone, you must soften the cavity by giving either black hellebore or wild purslane, mixing (with the hellebore) dauke, hartwort, cumin, anise or some other of the fragrant herbs, or (with the wild purslane) silphium juice. But in fact a mixture of these agents forms a complementary compound. The hellebore draws more forcefully and furthers the crisis better than the purslane, while the purslane expels wind downward better than the hellebore. Both of them relieve pain, as do many other purgatives, but these are the most effective ones I know of. Purgatives given in gruels are also good, unless they are excessively unpleasant either on account of their bitterness or some other taste, or in amount, or in color, or because they raise suspicion in the patient.

24. Immediately after the patient has taken the purgative potion, you should give him an amount of barley gruel to drink that is not significantly less than he is used to, although in general you should not give gruels during the cleaning itself; after the cleaning is finished, have him drink less gruel than usual. From then on the amount should be gradually increased, if the pain ceases and nothing else speaks against it.

25. My principle is the same if barley water must be given. I believe it is better in general to begin the administration at the (sc. disease's) onset, rather than, after having the patient first fast, giving the barley water on the third, fourth, fifth, sixth or seventh day, if the disease has

---

30 αἰεὶ MV: om. A

προκριθῇ ἡ νοῦσος ἐν τούτῳ τῷ χρόνῳ. αἱ δὲ προπαρασκευαὶ καὶ τούτοισι παραπλήσιοι ποιητέαι, ὁκοῖαί περ εἴρηνται.

26. (8 L.) Περὶ μὲν οὖν ῥυφήματος προσάρσιος οὕτω γινώσκω. ἀτὰρ καὶ περὶ ποτοῦ, ὁκοῖον ἄν τις μέλλῃ πίνειν, τῶν προσγραφησομένων ωὑτὸς μοι[31] λόγος τὸ ἐπίπαν ἐστίν. οἶδα δὲ τοὺς ἰητροὺς τὰ ἐναντιώτατα ἢ ὡς δεῖ ποιέοντας· βούλονται γὰρ ἅπαντες ὑπὸ τὰς ἀρχὰς τῶν νούσων προταριχεύσαντες τοὺς ἀνθρώπους ἢ δύο ἢ τρεῖς ἢ καὶ πλείους ἡμέρας οὕτω προσφέρειν τὰ ῥυφήματα καὶ τὰ ποτά· καὶ ἴσως τι καὶ εἰκὸς δοκέει αὐτοῖσι εἶναι μεγάλης τῆς μεταβολῆς γενομένης τῷ σώματι μέγα τι κάρτα καὶ ἀντιμεταβάλλειν. |

280 27. Τὸ δὲ μεταβάλλειν μὲν εὖ ἔχει μὴ ὀλίγον· ὀρθῶς μέντοι ποιητέη καὶ βεβαίως[32] ἡ μεταβολὴ καὶ ἔκ γε τῆς μεταβολῆς ἡ πρόσαρσις τῶν γευμάτων ἔτι μᾶλλον. μάλιστα μὲν οὖν ἂν βλάπτοιντο, εἰ μὴ ὀρθῶς μεταβάλλοιεν, οἱ οὔλῃσι τῇσι πτισάνῃσι χρεώμενοι· βλάπτοιντο δ᾽ ἄν καὶ οἱ μούνῳ τῷ ποτῷ χρεώμενοι, βλάπτοιντο δ᾽ ἂν καὶ οἱ μούνῳ τῷ χυλῷ χρεώμενοι, ἥκιστα δ᾽ ἂν οὗτοι.

28. (9 L.) Χρὴ δὲ καὶ τὰ μαθήματα ποιέεσθαι ἐν τῇ διαίτῃ τῶν ἀνθρώπων ἔτι ὑγιαινόντων, οἷα ξυμφέροι. 282 εἰ γὰρ δὴ τοῖσί γε | ὑγιαίνουσι φαίνεται διαφέροντα μεγάλα τὰ τοῖα ἢ τοῖα διαιτήματα καὶ ἐν ἄλλῳ τινὶ καὶ ἐν τῇσι μεταβολῇσι, πῶς οὐχὶ καὶ ἐν τῇσι νούσοισι διαφέρει μέγα καὶ τούτων ἐν τῇσιν ὀξυτάτῃσι

not reached its crisis by that time. The preparations for these patients, too, are to be made in the way indicated above.

26. With regard to the administration of gruels, this is what I think. As for drinks—how they should be taken—the underlying principle of what I am about to present will be generally the same. I am convinced, however, that physicians do exactly the opposite of what they should, for they all want at the diseases' onset to reduce the patient by having him fast for two, three, or more days, and only then to give the gruels and drinks—probably it seems somehow reasonable to them, given the violent change that is going on in the body, to counteract this change with some violent measure.

27. To bring about a change is no trivial benefit, but the change must be achieved properly and effectively, and the administration of foods after the change takes place must be even more correct. Now the patients most often harmed when the change is incorrectly made are those taking unstrained barley gruel, but the ones who are taking only drinks may also be harmed, and even those who employ only barley water, although these least.

28. It is also necessary to investigate, in the regimen of people who are still healthy, what things will be beneficial for them. For if in the healthy, greatly different effects result from one regimen or another, especially during changes, why would there not be significant differences during diseases, and the greatest during acute diseases?

---

[31] *μοι* MV: om. A

[32] *ποιητέη καὶ βεβαίως* A: *γε μεταβιβαστέη καὶ* MV

μέγιστα; ἀλλὰ μὴν εὐκαταμάθητόν γέ ἐστιν, ὅτι φαύλη δίαιτα βρώσιος καὶ πόσιος αὐτὴ ἑωυτῇ ἐμφερὴς αἰεὶ ἀσφαλεστέρη ἐστὶν τὸ ἐπίπαν ἐς ὑγιείην, ἢ εἴ τις ἐξαπίνης μέγα μεταβάλλοι ἐς ἄλλο.[33] ἐπεὶ καὶ τοῖσι δὶς σιτεομένοισι τῆς ἡμέρης καὶ τοῖσι μονοσιτέουσιν αἱ ἐξαπιναῖοι μεταβολαὶ βλάβας καὶ ἀρ-

284 ρωστίην παρέχουσιν. καὶ τοὺς μέν γε μὴ[34] | μεμαθηκότας ἀριστᾶν,[35] ἢν ἀριστήσωσιν, εὐθέως ἀρρώστους ποιέει καὶ βαρέας ὅλον τὸ σῶμα καὶ ἀσθενέας καὶ ὀκνηρούς· ἢν δὲ καὶ ἐπιδειπνήσωσιν, ὀξυρεγμιώδεας. ἐνίοισι δ' ἂν καὶ σπατίλη γένοιτο, ὅτι[36] παρὰ τὸ ἔθος ἠχθοφόρηκεν ἡ κοιλίη εἰθισμένη ἐπιξηραίνεσθαι καὶ μὴ δὶς διογκοῦσθαι μηδὲ δὶς ἕψειν τὰ σιτία.

286 29. Ἀρήγει οὖν τούτοισιν ἀνασηκῶσαι | τὴν μεταβολήν· ἐγκοιμηθῆναι γὰρ χρή, ὥσπερ νύκτα ἄγοντα μετὰ τὸ δεῖπνον, τοῦ μὲν χειμῶνος ἀρριγέως, τοῦ δὲ θέρεος ἀθαλπέως· ἢν δὲ καθεύδειν μὴ δύνηται, βραδεῖαν, συχνὴν ὁδὸν περιπλανηθέντα, μὴ στασίμως, δειπνῆσαι μηδὲν ἢ ὀλίγα μηδὲ βλαβερά· ἔτι δὲ ἔλασσον πιεῖν καὶ μὴ ὑδαρές. ἔτι δὲ μᾶλλον ἂν πονήσειεν ὁ τοιοῦτος, εἰ τρὶς σιτέοιτο[37] τῆς ἡμέρης ἐς κόρον· ἔτι

288 δὲ μᾶλλον, εἰ | πλεονάκις· καίτοι γε πολλοί εἰσιν οἳ εὐφόρως φέρουσι τρὶς σιτεόμενοι τῆς ἡμέρης ἐς πλῆθος, οἳ ἂν οὕτως ἐθισθῶσιν.

30. Ἀλλὰ μὴν καὶ οἱ μεμαθηκότες δὶς σιτέεσθαι τῆς

[33] Add. κρέσσον MV [34] μὴ A: om. MV

[35] ἀρισταν A: ἀναρισταν MV

But in fact it is not difficult to grasp that an ordinary regimen of food and drink by itself, consistently adhered to, is always safer for health, than if someone suddenly makes a great change to another regimen. Besides, for both those who ordinarily eat two meals a day and those who ordinarily eat one, major changes of this pattern bring injuries and diseases. In fact, those who are not in the habit of taking lunch, if they do so, will immediately become ill and heavy in their whole body, as well as weak and sluggish. If they then take dinner besides, they will suffer from acid eructations; in some even diarrhea results, because their cavity is bearing a burden against its habit, since it is used to having a period of dryness, and not being distended twice or concocting food twice.

29. It benefits these patients to compensate for any such change: they must take a nap, as someone would sleep after dinner, protecting themselves from cold in the winter, and from oppressive heat in the summer. If such a person cannot sleep, he should take a slow, regular walk with no rests, eat either nothing or very sparingly for dinner, and nothing that could be injurious. As drink, he should consume less yet, and nothing watery. A patient in this situation would suffer even more if he ate as much as he desired three times a day, and yet more again is he ate even oftener. Still there are many people who can happily stand eating hearty meals three times a day, since this is their habit.

30. On the other hand, people used to eating twice a

---

36 ὅτι A: ὁκόταν MV

37 σιτέοιτο MV: φάγοιεν A

ἡμέρης, ἢν μὴ ἀριστήσωσιν, ἀσθενέες καὶ ἄρρωστοί εἰσι καὶ δειλοὶ ἐς πᾶν ἔργον καὶ καρδιαλγέες· κρέμασθαι γὰρ αὐτοῖσι δοκέει τὰ σπλάγχνα, καὶ οὐρέουσι θερμὸν καὶ χλωρόν, καὶ ἡ ἄφοδος συγκαίεται. ἔστι δ' οἷσι καὶ τὸ στόμα πικραίνεται καὶ οἱ ὀφθαλμοὶ κοι-
290 λαίνονται καὶ οἱ κρόταφοι πάλλονται καὶ τὰ | ἄκρα διαψύχονται, καὶ οἱ μὲν πλεῖστοι τῶν ἀνηριστηκότων οὐ δύνανται κατεσθίειν τὸ δεῖπνον, δειπνήσαντες δὲ βαρύνουσι τὴν κοιλίην καὶ δυσκοιτέουσι πολὺ μᾶλλον ἢ εἰ προηριστήκεισαν.

31. Ὁκότε οὖν ταῦτα τοιαῦτα γίνεται τοῖσιν ὑγιαίνουσιν εἵνεκεν ἡμίσεος ἡμέρης διαίτης μεταβολῆς, παρὰ τὸ ἔθος οὔτε προσθεῖναι λυσιτελέειν φαίνεται οὔτε ἀφελέειν.

32. Εἰ τοίνυν οὗτος ὁ παρὰ τὸ ἔθος μονοσιτήσας ὅλην τὴν ἡμέρην κενεαγγήσας δειπνήσειεν ὁκόσον εἴθιστο, εἰκὸς αὐτόν, εἰ τότε ἀνάριστος ἐὼν ἐπόνεε καὶ
292 ἠρρώστει, δειπνήσας δὲ τότε βαρὺς ἦν, πολὺ | μᾶλλον βαρύνεσθαι· εἰ δέ γε ἔτι πλείω χρόνον κενεαγγήσας
294 | ἐξαπίνης μεταδειπνήσειεν, ἔτι μᾶλλον ἂν βαρύνοιτο.

33. Τὸν οὖν παρὰ τὸ ἔθος κενεαγγήσαντα ξυμφέρει ταύτην τὴν ἡμέρην ἀντισηκῶσαι ὧδε· ἀρριγέως καὶ ἀθαλπέως καὶ ἀταλαιπώρως—ταῦτα γὰρ πάντα βαρέως ἂν ἐνέγκοι—καὶ τὸ δεῖπνον συχνῷ ἔλασσον ποιήσασθαι ἢ ὅσον εἴθιστο καὶ μὴ ξηρόν, ἀλλὰ τοῦ πλαδαρωτέρου τρόπου· καὶ μετὰ ταῦτα πιεῖν μὴ ὑδαρὲς μηδὲ ἔλασσον ἢ κατὰ λόγον τοῦ βρώματος·

day, if they omit their first meal, become weak and sluggish, are adverse to any toil, and suffer from heartburn. They feel as if their inward parts are hanging down, their urine is hot and very yellow, and their stools are parched. In some cases, their mouth becomes bitter, their eyes are hollow, and their temples throb; the extremities are chilled. Most of those who have not taken their customary first meal are unable to get their dinner down either; if they do dine, their cavity is weighed down, and they sleep much worse than if they had taken their lunch.

31. Now since all these kinds of things happen to people in health, as a result of a half-day's change of regimen, it is clearly unprofitable for anyone either to augment or to subtract from their habitual custom.

32. If this person who took only one meal against his habit, then fasts again for a whole day, before eating his dinner in the same amount as usual, it is likely—since he once before, after going without his lunch, suffered pain and weakness, and then on dining felt heavy—that this time he will be weighed down even more; also that if he continues to practice this unaccustomed fasting for an even longer time, and then suddenly follows it with dinner, he will be weighed down yet more again.

33. Now the person who has fasted against his custom is benefited by compensating for the change of regimen on the same day in the following ways: to protect himself from any chill, oppressive heat, or exhaustion—all these would be difficult for him to tolerate—to take a considerably smaller dinner than is his habit, not dry, but of a moister kind; and after that to take neither too much drink nor less than corresponds with his food. On the next day

296 καὶ τῇ ὑστεραίῃ | ὀλίγα ἀριστῆσαι, ὡς ἐκ προσαγωγῆς ἀφίκηται ἐς τὸ ἔθος.

34. Αὐτοὶ μέντοι σφῶν αὐτῶν δυσφορώτερον δὴ τὰ τοιαῦτα φέρουσιν οἱ πικρόχολοι τὰ ἄνω· τὴν δὲ ἀσιτίην τὴν παρὰ τὸ ἔθος οἱ φλεγματίαι τὰ ἄνω εὐφορώτερον φέρουσι τὸ ἐπίπαν, ὥστε καὶ τὴν μονοσιτίην τὴν παρὰ τὸ ἔθος εὐφορώτερον[38] ἂν οὗτοι ἐνέγκαιεν.

35. Ἱκανὸν μὲν οὖν καὶ τοῦτο σημήϊον, ὅτι αἱ μέγισται μεταβολαὶ τῶν περὶ τὰς φύσιας ἡμέων καὶ τὰς ἕξιας ξυμβαινόντων μάλιστα νοσοποιέουσιν. οὐ δὴ οἷόν τε[39] παρὰ καιρὸν οὔτε σφοδρὰς τὰς κενεαγγίας ποιέειν οὔτε ἀκμαζόντων τῶν νοσημάτων καὶ ἐν φλεγμασίῃ ἐόντων προσφέρειν, οὔτε ἐξαπίνης οἷόν τε[39] ὅλῳ τῷ πρήγματι μεταβάλλειν οὔτε ἐπὶ τὰ οὔτε ἐπὶ τά. |

298 36. (10 L.) Πολλὰ δ' ἄν τις ἠδελφισμένα τούτοισι τῶν ἐς κοιλίην καὶ ἄλλα εἴποι, ὡς εὐφόρως μὲν φέρουσι τὰ βρώματα, ἃ εἰθίδαται,[40] ἢν καὶ μὴ ἀγαθὰ ᾖ φύσει· ὡσαύτως δὲ καὶ τὰ ποτά· δυσφόρως δὲ φέρουσι τὰ βρώματα, ἃ μὴ εἰθίδαται,[40] κἢν μὴ κακὰ ᾖ· ὡσαύτως δὲ καὶ τὰ ποτά.

37. Καὶ ὁκόσα μὲν κρεηφαγίη πολλὴ παρὰ τὸ ἔθος βρωθεῖσα ποιέει ἢ σκόροδα ἢ σίλφιον ἢ ὀπὸς ἢ καυλὸς ἢ ἄλλα ὁκόσα τοιουτότροπα μεγάλας δυνάμιας ἰδίας ἔχοντα, ἧσσον ἄν τις θαυμάσειεν, εἰ τὰ τοιαῦτα πόνους ἐμποιέει ἐν τῇσι κοιλίῃσι μᾶλλον τῶν ἄλλων.

[38] εὐφορώτερον AV: ξυμφερώτερον M

he should take a light lunch, in order that he will gradually return to his habitual regimen.

34. Such things are more difficult to tolerate for people who suffer from bilious humors in their upper (sc. cavity), whereas fasting against custom is generally better tolerated by people in whose upper cavity phlegm predominates, making it easier for them to endure having only one meal as an exception.

35. This, then, is sufficient proof that the most frequent causes of diseases are the violent changes that take place in our constitutions and our temporary states. For this reason, it is wrong to order fasts at the wrong times, or that are too strict, or when diseases are at their peak and in an inflammatory stage, or suddenly to change the whole regimen in one direction or another.

36. One could also add to this account many other related things regarding what enters the cavity, demonstrating that people tolerate foods to which they are accustomed even if these are not particularly good by nature—it is the same with drinks—while at the same time they do not tolerate foods against their habit even if these have no defect, and the same with drinks.

37. As to the effects caused by the consumption against a person's habit of much meat, or of garlic or silphium (whether juice or stalk), or of other substances like these with strong specific potencies, a person would be less likely to wonder that these produce more disruption in the cavity than other things do. But it is surprising to learn

---

39 *οἷόν τε* Gal: *οἴονται* codd.

40 *εἰθίδαται* A: *εἰθισμένοι εἰσί(ν)* MV

ἀλλὰ εἰ δὲ καταμάθοι, ὅσον μᾶζα ὄχλον καὶ ὄγκον
300 καὶ φῦσαν καὶ στρόφον | κοιλίῃ παρέχει παρὰ τὸ ἔθος βρωθεῖσα τῷ ἀρτοφαγεῖν[41] εἰθισμένῳ ἢ ὁκοῖον ἄρτος βάρος καὶ στάσιν κοιλίης τῷ μαζοφαγέειν εἰθισμένῳ[42] ἢ αὐτός γε ὁ ἄρτος θερμὸς βρωθεὶς οἵην δίψαν παρέχει καὶ ἐξαπιναίην πληθώρην διὰ τὸ ξηραντικόν τε καὶ βραδύπορον, καὶ οἱ ἄγαν καθαροί τε καὶ ξυγκομιστοὶ παρὰ τὸ ἔθος βρωθέντες οἷα διαφέροντα ἀλλήλων ποιέουσι καὶ μᾶζά γε ξηρὴ παρὰ τὸ ἔθος ἢ ὑγρὴ ἢ γλίσχρη, καὶ τὰ ἄλφιτα οἷόν τι ποιέει τὰ ποταίνια τοῖσι μὴ εἰωθόσι καὶ τὰ ἑτεροῖα τοῖσι τὰ ποταίνια εἰωθόσι· καὶ οἰνοποσίη καὶ ὑδροποσίη παρὰ τὸ ἔθος ἐς θάτερα μεταβληθέντα ἐξαπίνης καὶ ὑδαρής τε οἶνος καὶ ἄκρητος παρὰ τὸ ἔθος ἐξαπίνης ποθείς—ὁ μὲν γὰρ πλάδον τε ἐν τῇ ἄνω κοιλίῃ
302 ἐμποιήσει καὶ φῦσαν ἐν τῇ κάτω, ὁ δὲ παλμόν | τε φλεβῶν καὶ καρηβαρίην καὶ δίψαν—καὶ λευκός τε καὶ μέλας οἶνος[43] παρὰ τὸ ἔθος μεταβάλλουσιν, εἰ καὶ ἄμφω οἰνώδεες εἶεν, ὅμως πολλὰ ἂν ἑτεροιώσειαν κατὰ τὸ σῶμα· ὡς δὴ γλυκύν τε καὶ οἰνώδεα ἧσσον ἄν τις φαίη θαυμαστὸν εἶναι μὴ τωὐτὸ δύνασθαι ἐξαπίνης μεταβληθέντα.

(11 L.) Τιμωρητέον μὲν δὴ τοιόνδε τι μέρος τῷ ἐναντίῳ λόγῳ· ὅτι ἡ[44] μεταβολὴ τῆς διαίτης τούτοισιν ἐγένετο οὐ μεταβάλλοντος τοῦ σώματος οὔτ' ἐπὶ τὴν

[41] ἀρτοφαγεῖν A: μὴ μαζοφαγεῖν MV
[42] ἢ ὁκοῖον . . . εἰθισμένῳ M: om. AV

how much trouble, distension, flatulence, and colic barley cake produces in the cavity, when it is eaten against habit by a person used to eating bread. Or how much weight and stagnation arises in the cavity if bread is eaten by a person accustomed to eating barley cake. Or how much thirst bread by itself provokes, when it is eaten hot, together with sudden bloating, because of its drying and impeding character; and what different effects excessively refined or overcoarse bread causes when eaten against habit. And (sc. the effects which) barley cake that happens to be drier than normal, or moister, or overly viscid, and barley meal that is fresh has on people not used to it; or (sc. the effects) of the other kind of barley meal on people who are used to the fresh. And the effects of drinking wine and water against habit, if someone suddenly changes from the one to the other, or if wine is suddenly drunk diluted or unmixed against custom: the former causes fluid to collect in the upper cavity and flatulence in the lower cavity, while the latter causes throbbing of the vessels, heaviness in the head, and thirst. And to switch between white and dark wine against custom, even though both are dry, would likewise provoke many changes through the body. Hence a person would say it was less surprising that sweet and dry wines did not have the same effects if they were suddenly exchanged.

However, a certain concession might be made to the opposite argument, (sc. in recognizing) that in these examples the change of regimen was made without the body

43 οἶνος MV: om. A
44 ὅτι ἡ Gal: ὅτι μὴ M: om. A: ὅτι V

ῥώμην, ὥστε προσθέσθαι δεῖν σιτία, οὔτ᾿ ἐπὶ τὴν ἀρρωστίην, ὥστ᾿ ἀφαιρεθῆναι.

38. Προστεκμαρτέα δὴ καὶ ἡ ἰσχὺς καὶ ὁ τρόπος 304 τοῦ νοσήματος | ἑκάστου[45] καὶ τῆς φύσιος τοῦ [τε][46] ἀνθρώπου καὶ τοῦ ἔθεος τῆς διαίτης τοῦ κάμνοντος. οὐ μοῦνον σιτίων, ἀλλὰ καὶ ποτῶν. πολλῷ δ᾿ ἧσσον ἐπὶ τὴν πρόσθεσιν ἰτέον· ἐπεί[47] γε τὴν ἀφαίρεσιν ὅλως ἀφελεῖν πολλαχοῦ λυσιτελέει, ὅκου διαρκέειν μέλλει ὁ κάμνων, μέχρι ἂν τῆς νούσου ἡ ἀκμὴ πεπανθῇ. ἐν ὁκοίοισι δὲ τὸ τοιόνδε ποιητέον, γεγράψεται.

39. Πολλὰ δ᾿ ἄν τις καὶ ἄλλα ἠδελφισμένα τοῖσιν εἰρημένοισι γράφοι· τόδε γε μὴν κρέσσον μαρτύριον· οὐ γὰρ ἠδελφισμένον μοῦνόν ἐστι τῷ πρήγματι, περὶ οὗ μοι ὁ πλεῖστος λόγος εἴρηται, ἀλλ᾿ αὐτὸ τὸ πρῆγμα ἐπικαιρότατόν ἐστιν διδακτήριον. οἱ γὰρ ἀρχόμενοι τῶν ὀξέων νοσημάτων ἔστιν ὅτε οἱ μὲν σιτία ἔφαγον 306 | αὐθημερὸν ἠργμένοι ἤδη, οἱ δὲ καὶ τῇ ὑστεραίῃ, οἱ δὲ καὶ ἐρρύφεον τὸ προστυχόν, οἱ δὲ καὶ κυκεῶνα ἔπιον.[48] ἅπαντα δὲ ταῦτα κακίω μέν ἐστιν, ἢ εἰ ἑτεροίως[49] διαιτηθείη· πολλῷ μέντοι ἐλάσσω βλάβην φέρει ἐν τούτῳ τῷ χρόνῳ ἁμαρτηθέντα, ἢ εἰ τὰς μὲν πρώτας δύο ἡμέρας ἢ τρεῖς κενεαγγήσειε τελέως, τεταρταῖος δὲ ἐὼν τοιάδε διαιτηθείη ἢ πεμπταῖος· ἔτι

45 ἑκάστου A: ἑκάστοισι MV
46 τε codd.: del. Littré
47 ἐπεί V Gal: ἐπὶ MA

having any change either for the stronger, so that it required more food to be given, or in the direction of weakness, so that food had to be held back.[3]

38. To be heeded in addition are the strength and the nature of each disease, the constitution of the person, and the habitual regime of the patient, not only regarding foods, but also drinks. You should tend much less toward increasing portions, since on the contrary it is curtailing them completely that is often valuable, when a patient can afford to live on his reserves until the disease reaches its crisis with concoction. I will explain below in which circumstances this method should be applied.

39. A person could also add other things related to what has been said, but the following is a more convincing proof, for it is not just related to the subject which my account is mainly addressing, but is the actual subject itself, and a most vital lesson. At the beginning of their acute disease, people have sometimes already consumed food on that same day, others do so on the next day, others have taken whatever gruel happened to come to hand, and others have drunk a *cyceon*. All these actions are worse than if they had followed a different regimen, but these errors still bring much less injury at this time than if patients had fasted strictly on the first two or three days, and then taken up the incorrect regimen on the fourth or fifth day. Even

[3] How this sentence fits into the argument here is not at all clear, which may perhaps explain why the texts are so different in the three witnesses.

---

48 ἔπιον AV: ἐρρόφεον M 49 τις add. M

μέντοι κάκιον, εἰ ταύτας πάσας τὰς ἡμέρας προκενεαγγήσας ἐν τῇσιν ὕστερον ἡμέρῃσιν οὕτω διαιτηθείη, πρὶν ἢ πέπειρον τὴν νοῦσον γενέσθαι. οὕτω μὲν γὰρ θάνατον φέρει φανερῶς τοῖσι πλείστοισιν, εἰ μὴ πάμπαν ἡ νοῦσος εὐήθης εἴη. αἱ δὲ κατ᾽ ἀρχὰς ἁμαρτάδες οὐχ ὁμοίως ταύτῃσιν ἀνήκεστοί εἰσιν, ἀλλὰ πολλῷ εὐακεστότεραι. τοῦτο οὖν ἡγεῦμαι μέγιστον διδακτήριον, ὅτι οὐ στερητέαι αἱ πρῶται ἡμέραι τοῦ
308 ῥυφήματος ἢ τοίου ἢ τοίου | τοῖσι μέλλουσιν ὀλίγον ὕστερον ῥυφήμασιν ἢ σιτίοισι χρέεσθαι.

40. Πυθμενόθεν μὲν οὖν οὐκ ἴσασιν οὐθ᾽ οἱ τῇσι κριθώδεσι πτισάνῃσι χρεώμενοι, ὅτι αὐτῇσι[50] κακοῦνται, ὁκόταν ῥυφέειν ἄρξωνται, ἢν προκενεαγγήσωσιν δύο ἢ τρεῖς ἡμέρας ἢ πλείους, οὔτ᾽ αὖ οἱ τῷ χυλῷ χρεώμενοι γινώσκουσιν ὅτι τοιούτοισι βλάπτονται ῥυφέοντες, ὁκόταν μὴ ὀρθῶς ἄρξωνται τοῦ ῥυφήματος. τόδε γε μὴν καὶ φυλάσσουσι καὶ γινώσκουσιν, ὅτι μεγάλην τὴν βλάβην φέρει, ἤν, πρὶν πέπειρον τὴν νοῦσον γενέσθαι, κριθώδεα πτισάνην ῥυφῇ ὁ κάμνων, εἰθισμένος χυλῷ χρέεσθαι.

41. Πάντα οὖν ταῦτα μεγάλα μαρτύρια, ὅτι οὐκ
310 ὀρθῶς ἄγουσιν ἐς τὰ διαιτήματα | οἱ ἰητροὶ τοὺς κάμνοντας· ἀλλ᾽ ἐν ᾗσί τε νούσοισιν οὐ χρὴ κενεαγγέειν τοὺς μέλλοντας ῥυφήμασι διαιτᾶσθαι, κενεαγγέουσιν, ἐν ᾗσί τε οὐ χρὴ μεταβάλλειν ἐκ κενεαγγίης ἐς ῥυφήματα, ἐν ταύτῃσι μεταβάλλουσι. καὶ ὡς ἐπὶ

[50] -τῇσι Gal: -τοῖσι codd.

worse, though, would be if they first fasted on all these days, and then on the later days took up the regimen before the disease arrived at its concoction. Such missteps obviously bring death to the majority, unless the disease happens to be completely benign. Cases where the error was made at the onset are not as irremediable as these, but much easier to treat. Now I believe that this is the most important piece of evidence showing that the first days must not go without gruels, of one kind or another, in patients who are then going to employ gruels and foods a little later.

40. At bottom, then, there is a complete lack of understanding both among those taking unstrained gruels, who do not know that these cause damage if their use begins after two, three, or more days of fasting, and among those taking barley water, who do not know that they are injured by taking this, if they do not begin the draft at the correct time. The following, though, patients do pay attention to and understand: that it brings great harm if before a disease reaches concoction someone takes barley gruel, when their custom has been to take barley water.

41. All these, then, are convincing proofs that physicians do not guide their patients correctly in their regimen, but in diseases where patients about to take gruels should not be made to fast, they make them fast, and in those where there should be no change away from fasting to the administration of gruels, they make them change.

τὸ πολὺ ἀπαρτὶ[51] ἐν τούτοισι τοῖσι καιροῖσι μεταβάλ-
312 λουσιν ἐς | τὰ ῥυφήματα ἐκ τῆς κενεαγγίης, ἐν οἷσι πολλάκις ἀρήγει ἐκ τῶν ῥυφημάτων πλησιάζειν τῇ κενεαγγίῃ, ἢν οὕτω τύχῃ παροξυνομένη ἡ νοῦσος.

42. Ἐνίοτε δὲ καὶ ὠμὰ ἐπισπῶνται ἀπὸ τῆς κεφαλῆς καὶ τοῦ περὶ θώρηκα τόπου χολώδεα· ἀγρυπνίαι τε ξυνεμπίπτουσιν αὐτοῖσι, δι᾽ ἃς οὐ πέσσεται ἡ νοῦσος, περίλυποι δὲ καὶ πικροὶ γίνονται καὶ παραφρονέουσι, καὶ μαρμαρυγώδεά σφεων τὰ ὄμματα καὶ
314 αἱ ἀκοαὶ | ἤχου μεσταί· καὶ τὰ ἀκρωτήρια κατεψυγμένα καὶ οὖρα ἄπεπτα καὶ πτύσματα λεπτὰ καὶ ἁλυκὰ καὶ κεχρωσμένα ἀκρήτῳ χρώματι σμικρὰ καὶ ἱδρῶτες περὶ τράχηλον καὶ διαπορήματα, καὶ πνεῦμα προσπταῖον ἐν τῇ ἄνω φορῇ πυκνὸν ἢ μέγα λίην, ὀφρύες δεινώσιος μετέχουσαι, λειποψυχώδεα πονηρὰ καὶ τῶν ἱματίων ἀπορρίψιες ἀπὸ τοῦ στήθεος καὶ χεῖρες τρομώδεες, ἐνίοτε δὲ καὶ χεῖλος τὸ κάτω σείεται. ταῦτα δ᾽ ἐν ἀρχῇσι παραφαινόμενα παραφροσύνης δηλωτικά ἐστι σφοδρῆς, καὶ ὡς ἐπὶ τὸ πολὺ ἀποθνήσκουσιν· οἱ δὲ διαφεύγοντες ἢ μετὰ ἀποστήματος ἢ αἵματος ῥύσιος ἐκ τῆς ῥινὸς ἢ πῦον παχὺ πτύσαντες, ἄλλως δὲ οὔ.

316 43. Οὐδὲ γὰρ τῶν τοιούτων ὁρῶ ἐμπείρους | τοὺς ἰητρούς, ὡς χρὴ διαγινώσκειν τὰς ἀσθενείας ἐν τῇσι νούσοισιν, αἵ τε διὰ κενεαγγίην ἀσθενέονται, αἵ τε δι᾽ ἄλλον τινὰ ἐρεθισμόν, αἵ τε διὰ πόνον καὶ ὑπὸ ὀξύτη-

And in most cases they make the change from fasting to gruel at just those crucial moments when it often helps to move from gruels toward fasting, namely when the disease happens to be in a state of exacerbation.

42. Sometimes (sc. such treatment) also draws unconcocted material from the head, and bilious material from the area of the thorax; sleeplessness may arise in such patients, which prevents the disease from being concocted: the patients become very depressed, angry, and delirious, their eyes see flashes of light, and their ears are filled with echoing; their extremities become cold, their urines remain unconcocted, their sputa are thin, salty, and slightly suffused with a uniform color, and sweats break out around their necks in conjunction with restlessness; their breath is held back in its upward movement, being rapid and very deep; the eyebrows frown, there is an ominous loss of consciousness, they tear the bedding away from their chest, their hands tremble, and sometimes the lower lip quivers. For these signs to appear in the early stages indicates an incipient severe delirium, and usually such patients succumb, others surviving by having an abscession, a hemorrhage from the nose, or coughing up thick pus—but not otherwise.

43. Nor indeed do I see that physicians are as competent as they should be at recognizing the kinds of weakness that occur in diseases—some diseases induce weakness due to starvation, others due to some other kind of irritation, and others because of the pain and acuteness of the

---

[51] *ἀπαρτὶ* Joly: *ἁμαρτάνουσιν. ἐνίοτε δὲ* M: *ἁμαρτάνει ὅτε δ'* A: *ἁμαρτάνουσιν ἐνίοτε* V: *ἀπαρτὶ ἐνίοτε* Gal

τος τῆς νούσου, ὁκόσα τε ἡμέων ἡ φύσις καὶ ἡ ἕξις ἑκάστοισιν ἐκτεκνοῖ πάθεα καὶ εἴδεα παντοῖα· καίτοι σωτηρίην ἢ θάνατον φέρει γινωσκόμενα ἢ ἀγνοεύμενα τὰ τοιαῦτα.

44. Μέζον μὲν γὰρ κακόν ἐστιν, ἢν διὰ τὸν πόνον καὶ τὴν ὀξύτητα τῆς νούσου ἀσθενέοντι προσφέρῃ τις ποτὸν ἢ ῥύφημα πλέον ἢ σιτίον, οἰόμενος διὰ κενεαγγίην ἀσθενέειν. ἀεικὲς δὲ καὶ διὰ κενεαγγίην ἀσθενέοντα μὴ γνῶναι καὶ πιέζειν τῇ διαίτῃ· φέρει μὲν γάρ τινα κίνδυνον καὶ αὕτη ἡ ἁμαρτάς, πολλῷ δὲ ἧσσον τῆς ἑτέρης· καταγελαστοτέρη δὲ πολλῷ αὕτη μᾶλλον ἡ ἁμαρτὰς τῆς ἑτέρης· εἰ γὰρ ἄλλος ἰητρὸς ἢ καὶ 318 δημότης[52] | ἐσελθὼν καὶ γνούς τὰ συμβεβηκότα δοίη καὶ φαγεῖν καὶ πιεῖν, ἃ ὁ ἕτερος ἐκώλυεν, ἐπιδήλως ἂν δοκέοι ὠφεληκέναι. τὰ δὲ τοιάδε μάλιστα καθυβρίζεται τῶν χειρωνακτέων ὑπὸ τῶν ἀνθρώπων· δοκέει γὰρ αὐτοῖσιν ὁ ἐπεσελθὼν[53] ἰητρὸς ἢ ἰδιώτης ὡσπερεὶ τεθνεῶτα ἀναστῆσαι. γεγράψεται οὖν καὶ περὶ τούτων σημήϊα, οἷσι δεῖ[54] ἕκαστα τούτων διαγινώσκειν.

45. (12 L.) Παραπλήσια μέντοι τοῖσι κατὰ κοιλίην ἐστὶ καὶ ταῦτα· καὶ γὰρ ἢν ὅλον τὸ σῶμα ἀναπαύσηται πολὺ παρὰ τὸ ἔθος, οὐκ αὐτίκα ἔρρωται μᾶλλον· ἢν δὲ δὴ καὶ πλείω χρόνον διελινύσας ἐξαπίνης ἐς τοὺς πόνους ἔλθῃ, φλαῦρόν[55] τι πρήξει ἐπιδήλως. οὕτω δὲ καὶ ἓν ἕκαστον τοῦ σώματος· καὶ γὰρ οἱ

[52] δημότης A: ἰδιώτης MV [53] ἐπεσελθῶν A: ἐσελθῶν MV [54] δεῖ MV: χρὴ A [55] φλαῦρον MV: φαῦλόν A

disease—nor the affections and all their varieties that our nature and the temporary state (sc. of our body) engender in each of us. And yet recovery or death results respectively from understanding or failure to understand such things.

44. For example, it is a great mistake if, when a patient is weak because of the pain and acuteness of his disease, someone administers a drink, or more gruel or food because he (sc. mistakenly) believes that the weakness is the result of starvation. It is also blameworthy not to recognize when a person *is* weak from starvation, and to worsen his state by (sc. an incorrect) regimen; for this mistake too brings a definite danger, although much less than the previous one. But this mistake is much more likely to make the physician a laughingstock than the other one, since if another physician or even a layman comes in, and, recognizing the situation, gives the food and drink that the first one forbade, the help he has brought will be obvious. Such a thing especially causes practitioners to be mocked by the laity, who think that the second physician or layman has done something equivalent to raising the dead. Thus I will also give an account of the signs by which each of these conditions can be recognized.

45. Analogous to these considerations concerning the cavity (sc. which I have detailed) are the following. If the whole body at some time takes a complete rest, to which it is not accustomed, it will not immediately improve in health; and if after (sc. a person) has been completely idle for a longer time, he suddenly begins to labor, he will evidently fare quite badly. It is also similar for the individual parts of the body, for the feet and the other limbs will

320 πόδες τοιόνδε τι πρήξειαν καὶ τἆλλα ἄρθρα, | μὴ εἰθισμένα πονέειν, ἢν διὰ χρόνου ἐξαπίνης ἐς τὸ πονέειν ἔλθῃ· ταὐτὰ δ' ἂν καὶ οἱ ὀδόντες καὶ οἱ ὀφθαλμοὶ πάθοιεν, καὶ οὐδὲν ὅ τι οὔ· ἐπεὶ καὶ κοίτη παρὰ τὸ ἔθος μαλθακὴ πόνον ἐμποιέει καὶ σκληρὴ παρὰ τὸ ἔθος, καὶ ὑπαίθριος εὐνὴ[56] παρὰ τὸ ἔθος σκληρύνει τὸ σῶμα.

46. Ἀτὰρ τῶν τοιῶνδε πάντων ἀρκέει παράδειγμά τι γράψαι· εἰ γάρ τις ἕλκος λαβὼν ἐν κνήμῃ μήτε λίην ἐπίκαιρον μήτε λίην εὔηθες, μήτε ἄγαν εὐελκὴς ἐὼν μήτε ἄγαν δυσελκής, αὐτίκα ἀρξάμενος ἐκ πρώ-
322 της κατακείμενος ἰητρεύοιτο καὶ | μηδαμῇ μετεωρίζοι τὸ σκέλος, ἀφλέγμαντος μὲν ἂν εἴη οὗτος μᾶλλον καὶ ὑγιὴς πολλῷ θᾶσσον ἂν γένοιτο, ἢ εἰ πλανώμενος ἰητρεύοιτο. εἰ μέντοι πεμπταῖος ἢ ἑκταῖος ἐών, ἢ καὶ ἔτι ἀνωτέρω, ἀναστὰς ἐθέλοι προβαίνειν, μᾶλλον ἂν πονέοι τότε, ἢ εἰ αὐτίκα ἐξ ἀρχῆς πλανώμενος ἰητρεύοιτο. εἰ δὲ καὶ πολλὰ ταλαιπωρήσειεν ἐξαπίνης, πολλῷ ἂν μᾶλλον πονήσειεν, ἢ εἰ κείνως[57] ἰητρευόμε-
324 νος | τὰ αὐτὰ ταῦτα ταλαιπωρήσειεν ἐν ταύτῃσι τῇσιν ἡμέρῃσι. διὰ τέλεος οὖν μαρτυρεῖ ταῦτα πάντα ἀλλήλοισιν, ὅτι πάντα ἐξαπίνης μέζον πολλῷ τοῦ μετρίου μεταβαλλόμενα καὶ ἐπὶ τὰ καὶ ἐπὶ τὰ βλάπτει.

47. Πολλαπλασίη μὲν οὖν κατὰ κοιλίην ἡ βλάβη ἐστίν, ἢν ἐκ πολλῆς κενεαγγίης ἐξαπίνης πλέον τοῦ

[56] εὐνὴ MV: κοίτη A
[57] εἰ κείνως Littré: ἐκείνως M: κείνως A: κεῖνος V

suffer in this way if, when unaccustomed to labor for a long time, they suddenly set to work. The same would also befall the teeth and the eyes, and in fact there is no part of the body that would not suffer in the same way. And likewise, both a soft bed or a firm bed, if used against habit, will cause fatigue, just as the body of a person sleeping in the open air when this is not his habit will suffer from stiffness.

46. But regarding all these subjects, it suffices to record one example. If a person developed a lesion in his lower leg that was neither very critical nor completely harmless, being a person who tended generally neither to heal especially easily nor with particular difficulty, and he went to bed and began treatment on the first day, and he never lifted the leg up, he would be more likely to escape inflammation and to recover much more quickly than if, while he was being treated, he continued to move about. If however, on the fifth or sixth day, or even sooner, he wanted to get up and go outside, he would suffer more pain at that time, than if his treatment had been conducted right from the beginning with him moving about. And if he began suddenly to exert himself actively, he would suffer more pain than if, when treated the other way (i.e., still moving about), he had made the same exertions on the same days. All these cases, then, confirm one another in demonstrating that all sudden changes that go much beyond what is moderate, in one direction or the other, do harm.

47. Likewise, there will be many times more damage to the cavity, if, after long fasting, more than a moderate

326 μετρίου προσαίρηται—καὶ κατὰ τὸ | ἄλλο σῶμα, ἢν ἐκ πολλῆς ἡσυχίης ἐξαίφνης ἐς πλείω πόνον ἔλθῃ, πολλῷ πλείω βλάψει—ἢ εἰ ἐκ πολλῆς ἐδωδῆς ἐς κενεαγγίην μεταβάλλοι. δεῖ μέντοι καὶ τὸ σῶμα τούτοι-
328 σιν | ἐλινύειν. κἢν ἐκ πολλῆς ταλαιπωρίης ἐξαπίνης ἐς σχολήν τε καὶ ῥᾳθυμίην ἐμπέσῃ, δεῖ δὲ καὶ τούτοισι τὴν κοιλίην ἐλινύειν ἐκ πλήθεος βρώμης· εἰ δὲ μή, πόνον ἐν τῷ σώματι ἐμποιήσει καὶ βάρος ὅλου τοῦ σώματος.

48. (13 L.) Ὁ οὖν πλεῖστός μοι λόγος γέγονε περὶ τῆς μεταβολῆς τῆς ἐπὶ τὰ καὶ ἐπὶ τά. ἐς πάντα μὲν
330 οὖν εὔχρηστον | ταῦτ᾽ εἰδέναι· ἀτὰρ καί, περὶ οὗ ὁ λόγος ἦν, ὅτι ἐν τῇσιν ὀξείῃσι νούσοισιν ἐς τὰ ῥυφήματα μεταβάλλουσιν ἐκ τῆς κενεαγγίης· μεταβλητέον γὰρ ὡς ἐγὼ κελεύω· ἔπειτα οὐ χρηστέον ῥυφήμασι, πρὶν ἡ νοῦσος πεπανθῇ ἢ ἄλλο τι σημεῖον φανῇ ἢ κατὰ ἔντερον, κενεαγγικὸν ἢ ἐρεθιστικόν, ἢ κατὰ τὰ ὑποχόνδρια, ὁκοῖα γεγράψεται.[58]

49. Ἀγρυπνίη ἰσχυρὴ πόμα καὶ σιτία ὠμὰ καὶ[59]
332 ἀπεπτότερα ποιέει, καὶ ἡ ἐπὶ | θάτερα αὖ μεταβολὴ λύει τὸ σῶμα καὶ ἑφθότητα καὶ καρηβαρίην ἐμποιέει.

50. (14 L.) Γλυκὺν δὲ οἶνον καὶ οἰνώδεα, καὶ λευκὸν καὶ μέλανα, καὶ μελίκρητον καὶ ὕδωρ καὶ ὀξύμελι τοισίδε σημαινόμενον χρὴ διορίζειν ἐν τῇσιν ὀξείῃσι νούσοισι· ὁ μὲν γλυκὺς ἧσσόν ἐστιν καρηβαρικὸς τοῦ οἰνώδεος καὶ ἧσσον φρενῶν ἁπτόμενος καὶ διαχωρητικώτερος δή τι τοῦ ἑτέρου κατὰ ἔντερον, μεγα-

amount of food is suddenly taken—and more damage to the rest of the body, too, if from complete inaction there is a sudden change to active toil—than if from hearty eating there is a change to fasting. However, in these cases too the body must rest. But if the body suddenly falls from great exertion into leisure and ease, the cavity must also be spared from a surfeit of food: otherwise, there will be pain in the body and heaviness throughout the whole frame.

48. My account has dealt mainly with change in one direction or in the other. This is useful to understand in all cases, but in particular—and this is the center of my attention—because in acute diseases patients change from fasting to taking gruels. This change must be made in the manner I am recommending, since gruels should not be employed before the disease reaches a state of concoction, or some other sign appears, either of inanition, or of irritation in the intestine or hypochondrium, as I will describe below.

49. Persistent sleeplessness makes food and drink remain raw and more difficult to digest, while a change in the opposite direction makes the body slack, and provokes languor and heaviness in the head.

50. The use of wine—both sweet and dry, both white and dark—melicrat, water, and oxymel must be differentiated according to the following criteria in acute diseases. Sweet wine leads less often to heaviness in the head than dry wine, affects the mind less, is somewhat more laxative than the other one for the intestine, and causes swelling

---

[58] Add. σημεῖα MV [59] σιτία ὠμὰ καὶ MV: σιτίον A

λόσπλαγχνος δὲ σπληνὸς καὶ ἥπατος· οὐκ ἐπιτήδειος δὲ οὐδὲ τοῖσι πικροχόλοισι· καὶ γὰρ οὖν διψώδης τοῖσί γε τοιούτοις· ἀτὰρ καὶ φυσώδης τοῦ ἐντέρου τοῦ ἄνω, οὐ μὴν πολέμιός γε τῷ ἐντέρῳ τῷ κάτω ὡς κατὰ λόγον τῆς φύσης· καίτοι οὐ πάνυ πορίμη ἐστὶν ἡ ἀπὸ τοῦ γλυκέος οἴνου φῦσα, ἀλλ᾽ ἐγχρονίζει περὶ ὑπο-
334 χόνδρια. καὶ γὰρ οὖν | οὗτος ἧσσον διουρητικός ἐστιν τὸ ἐπίπαν τοῦ οἰνώδεος λευκοῦ· πτυάλου δὲ μᾶλλον ἀναγωγὸς τοῦ ἑτέρου ὁ γλυκύς. καὶ οἷσι μὲν διψώδης ἐστὶ πινόμενος, ἧσσον ἂν τούτοισιν ἀνάγοι ἢ ὁ ἕτερος οἶνος, οἷσι δὲ μὴ διψώδης, μᾶλλον ἀνάγοι ἂν τοῦ ἑτέρου.

51. Ὁ δὲ λευκὸς οἰνώδης οἶνος ἐπῄνηται μὲν καὶ ἔψεκται τὰ πλεῖστα καὶ τὰ μέγιστα ἤδη ἐν τῇ τοῦ γλυκέος οἴνου διηγήσει· ἐς δὲ κύστιν μᾶλλον πόριμος ἐὼν τοῦ ἑτέρου καὶ διουρητικὸς καὶ καταρρηκτικὸς ἐὼν αἰεὶ πολλὰ προσωφελέοι ἂν ἐν ταύτῃσι τῇσι νούσοισι· καὶ γὰρ εἰ πρὸς ἄλλα ἀνεπιτηδειότερος τοῦ ἑτέρου πέφυκεν, ἀλλ᾽ ὅμως κατὰ κύστιν ἡ κάθαρσις ὑπ᾽ αὐτοῦ γινομένη ῥύεται, ἢν προτρέπηται ὁκοῖον δεῖ. καλὰ δὲ ταῦτα τεκμήρια περὶ τοῦ οἴνου καὶ ὠφελείης καὶ βλάβης· ἄσσα ἀκαταμάθητα ἦν τοῖσιν ἐμεῦ γεραιτέροισι.

52. Κιρρῷ δ᾽ οἴνῳ καὶ μέλανι αὐστηρῷ ἐν ταύτῃσι τῇσι νούσοισιν ἐς τάδε ἂν χρήσαιο· εἰ καρηβαρίη |
336 μὲν μὴ ἐνείη μηδὲ φρενῶν ἅψις μηδὲ τὸ πτύαλον κωλύοιτο τῆς ἀνόδου μηδὲ τὸ οὖρον ἴσχοιτο, τὰ διαχωρήματα δὲ πλαδαρώτερα καὶ ξυσματωδέστερα εἴη, ἐν

of the spleen and liver. It is not suitable for patients with bitter bile, since it provokes thirst in them. Furthermore, sweet wine produces wind in the upper cavity, although it does not injure the lower intestine in proportion to the wind generated; actually, the wind coming from sweet wine is not very mobile, but remains for some time in the region of the hypochondrium. Sweet wine is generally less diuretic than dry white wine, but it does promote the expectoration of sputum more effectively than the other one. In patients who on drinking it become thirsty, it brings up less sputum than dry wine, while in patients with no thirst it brings up more sputum than the other one.

51. The most important advantages and disadvantages of dry white wine have already been presented in detail in my account of sweet wine. Since it flows more readily to the bladder than the other one, is diuretic, and promotes excretion, it will always have a use in acute diseases. For although it is in various other respects naturally less suited than sweet wine, nevertheless the cleaning through the bladder that it effects, when it is administered as it should be, gives relief. These are important testimonies to wine's benefit and harm, which have not been appreciated by my predecessors.

52. Tawny wine and dry dark wine can be prescribed in these diseases under the following circumstances. If there is no heaviness in the head, no affection of the mind, the sputum is not being prevented from coming up, the urine is not suppressed, and the stools are quite moist and

δὴ τοῖσι τοιούτοισι πρέποι ἂν μάλιστα μεταβάλλειν ἐκ τοῦ λευκοῦ καὶ ὁκόσα τούτοισιν ἐμφερέα. προσξυνιέναι δὲ δεῖ[60] *ὅτι τὰ μὲν ἄνω πάντα καὶ τὰ κατὰ κύστιν ἧσσον βλάψει, ἢν ὑδαρέστερος ᾖ, τὰ δὲ κατ' ἔντερον μᾶλλον ὀνήσει, ἢν ἀκρητέστερος ᾖ.*

53. (15 L.) *Μελίκρητον δὲ πινόμενον διὰ πάσης τῆς νούσου ἐν τῇσιν ὀξείῃσι νούσοισι τὸ ἐπίπαν μὲν τοῖσι πικροχόλοισι καὶ μεγαλοσπλάγχνοισιν ἧσσον ἐπιτήδειον ἢ τοῖσι μὴ τοιούτοισι· διψῶδές γε μὴν ἧσσον τοῦ γλυκέος οἴνου· πνεύμονός τε γὰρ μαλθακτικόν ἐστι καὶ πτυάλου ἀναγωγὸν μετρίως καὶ βηχὸς*
338 *παρηγορικόν· ἔχει γὰρ σμηγματῶδές* | *τι, ὃ* <*οὐ*>[61] *μᾶλλον τοῦ καιροῦ καταγλισχραίνει τὸ πτύαλον.* |
340 *ἔστι δὲ καὶ διουρητικὸν μελίκρητον ἱκανῶς, ἢν μή τι τῶν ἀπὸ σπλάγχνων κωλύῃ· καὶ διαχωρητικὸν δὲ κάτω χολωδέων, ἔστι μὲν ὅτε καλῶν, ἔστι δ' ὅτε κατα-*
342 *κορεστέρων* | *μᾶλλον τοῦ καιροῦ καὶ ἀφρωδεστέρων. μᾶλλον δὲ τὸ τοιοῦτο τοῖσι χολώδεσί τε καὶ μεγαλοσπλάγχνοισι γίνεται.*

54. *Πτυάλου μὲν οὖν ἀναγωγὴν καὶ πνεύμονος μάλθαξιν τὸ ὑδαρέστερον μελίκρητον ποιέει μᾶλλον· τὰ μέντοι ἀφρώδεα διαχωρήματα καὶ μᾶλλον τοῦ καιροῦ κατακορέως χολώδεα καὶ μᾶλλον θερμὰ τὸ ἄκρητον μᾶλλον τοῦ ὑδαρέος ἄγει· τὸ δὲ τοιόνδε διαχώρημα ἔχει μὲν καὶ ἄλλα σίνεα μεγάλα· οὔτε γὰρ ἐξ ὑποχονδρίων καῦμα σβεννύει, ἀλλ' ὁρμᾷ, δυσφορίην τε καὶ ῥιπτασμὸν τῶν μελέων ποιέει ἑλκῶδές τ' ἐστὶ καὶ ἐντέρου καὶ ἕδρης· ἀλεξητήρια δὲ τούτων γεγράψεται.*

contain shreds of flesh, in these and similar circumstances it is fitting to move away from white wine and all similar kinds. It must be understood, that these wines will do less damage in all the upper parts and in the region of the bladder if they are thinner, and will benefit the intestine more if they are purer.

53. When melicrat is drunk through the course of acute diseases, it is generally less suitable for bilious patients and those with enlarged viscera than for others. However, it provokes less thirst than sweet wine, softens the lung, moderately stimulates expectoration, and relieves coughing. This is because it has a certain detergent quality that prevents sputum from becoming too viscid. Melicrat is also sufficiently diuretic, unless some condition of the viscera prevents this effect. It promotes the excretion of bilious stools, which is sometimes favorable, but at other times they are too intense in color and foamier than they should be; this tends to happen more often in bilious patients and those with enlarged viscera.

54. More diluted melicrat tends to favor the expectoration of sputum and softening of the lung, while undiluted melicrat makes the stools foamy, more intensely bilious than normal, and hotter than diluted melicrat does. Such stools are also dangerous in other ways since they not only fail to extinguish the heat kindled in the hypochondrium, but actually increase it, besides causing discomfort and agitation of the limbs, and a tendency to ulceration in the intestine and seat. Ways to guard against these effects will be described below.

---

[60] δεῖ MV: χρὴ A [61] Add. Coray (in Joly)

55. Ἄνευ μὲν οὖν ῥυφημάτων μελικρήτῳ χρεώμενος ἀντ᾽ ἄλλου ποτοῦ ἐν ταύτῃσι τῇσι νούσοισι πολλὰ ἂν εὐτυχοίης καὶ οὐκ ἂν πολλὰ ἀτυχοίης· οἷσι δὲ δοτέον καὶ οἷσιν οὐ δοτέον, τὰ μέγιστα εἴρηται, καὶ δι᾽ ἃ οὐ δοτέον. |

344 56. Κατέγνωσται δὲ μελίκρητον ὑπὸ τῶν ἀνθρώπων, ὡς καταγυιοῖ τοὺς πίνοντας, καὶ διὰ τοῦτο ταχυθάνατον εἶναι νενόμισται. ἐκλήθη δὲ τοῦτο διὰ τοὺς ἀποκαρτερέοντας· ἔνιοι γὰρ μελικρήτῳ ποτῷ χρέονται ὡς τοιοῦδε δῆθεν ἐόντος. τὸ δὲ οὐ παντάπασιν ὧδε ἔχει, ἀλλὰ ὕδατος μὲν πολλῷ ἰσχυρότερόν ἐστι πινόμενον μοῦνον, εἰ μὴ ταράσσοι τὴν κοιλίην· ἀτὰρ καὶ οἴνου[62] λεπτοῦ καὶ ὀλιγοφόρου καὶ ἀνόσμου ᾗ μὲν

346 ἰσχυρότερον, ᾗ δὲ ἀσθενέστερον. | μέγα μὴν διαφέρει καὶ οἴνου καὶ μέλιτος ἀκρητότης ἐς ἰσχύν· ἀμφοῖν δ᾽ ὅμως τούτων, εἰ καὶ διπλάσιον μέτρον οἴνου ἀκρήτου πίνοι τις, ἢ ὁκόσον μέλι ἐκλείχοι, πολλῷ ἂν δήπου ἰσχυρότερος εἴη ὑπὸ τοῦ μέλιτος, εἰ μοῦνον μὴ ταράσσοι τὴν κοιλίην· πολλαπλάσιον γὰρ καὶ τὸ κόπριον διεξίοι ἂν αὐτῷ. εἰ μέντοι ῥυφήματι χρέοιτο πτισάνῃ, ἐπιπίνοι δὲ μελίκρητον, ἄγαν πλησμονῶδες ἂν εἴη καὶ φυσῶδες καὶ τοῖσι κατὰ τὰ ὑποχόνδρια σπλάγχνοισιν ἀξύμφορον· προπινόμενον μέντοι πρὸ ῥυφημάτων μελίκρητον οὐ βλάπτει ὡς μεταπινόμενον, ἀλλά τι καὶ ὠφελέει.

57. Ἑφθὸν δὲ μελίκρητον ἐσιδεῖν μὲν πολλῷ κάλλιον τοῦ ὠμοῦ· λαμπρὸν γὰρ καὶ λεπτὸν καὶ λευκὸν καὶ διαφανὲς γίνεται. ἀρετὴν δὲ ἥντινα αὐτῷ προσθέω

55. Now if in acute diseases you employ melicrat as a replacement for drink, but no gruels, you will frequently be successful and rarely fail. Which patients are suited for this treatment, and which not, has been discussed above, as well as why.

56. Melicrat has been condemned by people as enfeebling to those who drink it, and for this reason tending to accelerate death. But in fact it has received this (sc. false) reputation because some people who are starving themselves to death drink melicrat, believing that it will hasten their end. Actually, when it is drunk alone, it gives much more strength than water does, except if it provokes diarrhea. Furthermore, sometimes it has more strength in it than thin, meager, odorless wine does, although at other times it has less. Completely unmixed, wine and honey each excels differently in giving strength. As far as these two are concerned, if someone were to take twice as much unmixed wine as honey, he would still be strengthened much more by the honey, but only if the honey does not bring on diarrhea, since then he would also excrete many times as much. But if he took barley gruel before drinking the melicrat, he would feel excessively full, flatulent, and stirred up in his viscera around the hypochondrium: if he drank melicrat before the gruels, however, he would not be harmed as he would be by taking them in the other sequence, but even to a degree benefited.

57. Boiled melicrat is much more pleasant to look at than when it is raw, since (sc. by boiling) it becomes bright, thin, white, and transparent; however, I do not attribute

[62] Add. λευκοῦ καὶ M

διαφέρουσάν τι τοῦ ὠμοῦ οὐκ ἔχω· οὐδὲ γὰρ ἥδιόν[63]
348 ἐστιν τοῦ ὠμοῦ, ἢν τυγχάνῃ | γε τὸ μέλι καλὸν ἐόν·
ἀσθενέστερον μέντοι γε τοῦ ὠμοῦ καὶ ἀκοπρωδέστερόν ἐστιν· ὧν οὐδετέρης τιμωρίης προσδέεται μελίκρητον. ἄγχιστα δὲ χρηστέον αὐτῷ τοιῷδε ἐόντι, ἢν τὸ μέλι τυγχάνῃ πονηρὸν ἐὸν καὶ ἀκάθαρτον καὶ μέλαν καὶ μὴ εὐῶδες· ἀφέλοιτο γὰρ ἂν ἡ ἕψησις τῶν κακοτήτων αὐτοῦ τὸ πλεῖον τοῦ αἴσχεος.

58. (16 L.) Τὸ δὲ ὀξύμελι καλεόμενον ποτὸν πολλαχοῦ εὔχρηστον ἐν ταύτῃσι τῇσι νούσοισιν εὑρήσεις ἐόν· καὶ γὰρ πτυάλου ἀναγωγόν ἐστι καὶ εὔπνοον. καιροὺς μέντοι τοιούσδε ἔχει· τὸ μὲν κάρτα ὀξὺ οὐδὲν
350 ἂν μέσον | ποιήσειεν πρὸς τὰ πτύαλα τὰ μὴ ῥηϊδίως ἀνιόντα· εἰ γὰρ ἀναγάγοι μὲν τὰ ἐγκέρχνοντα καὶ ὄλισθον ἐμποιήσειε καὶ ὥσπερ διαπτερώσειε τὸν βρόγχον, παρηγορήσειεν ἄν τι τὸν πλεύμονα· μαλθακτικὸν γάρ. καὶ εἰ μὲν ταῦτα συγκυρήσειε, μεγάλην ὠφελείην ἐμποιήσει. ἔστι δ' ὅτε τὸ κάρτα ὀξὺ οὐκ ἐκράτησε τῆς ἀναγωγῆς τοῦ πτυάλου, ἀλλὰ προσεγλίσχρηνε καὶ ἔβλαψε· μάλιστα δὲ τοῦτο πάσχουσιν
352 οἵπερ καὶ ἄλλως ὀλέθριοί εἰσι καὶ | ἀδύνατοι βήσσειν τε καὶ ἀποχρέμπτεσθαι τὰ ἐνεχόμενα. ἐς μὲν οὖν τόδε προστεκμαίρεσθαι χρὴ τὴν ῥώμην τοῦ ἀνθρώπου καί, ἢν ἐλπίδα ἔχῃ, διδόναι· διδόναι δέ, ἢν διδῷς, ἀκροχλίερον καὶ κατ' ὀλίγον τὸ τοιόνδε καὶ μὴ λάβρως.

59. Τὸ μέντοι ὀλίγον ὕποξυ[64] ὑγραίνει μὲν τὸ στόμα καὶ φάρυγγα ἀναγωγὸν δὲ πτυάλου ἐστὶ καὶ ἄδιψον· ὑποχονδρίῳ δὲ καὶ σπλάγχνοισιν τοῖσι ταύτῃ εὐ-

any additional effectiveness to the boiled that is absent in the raw; the boiled is also no sweeter than the raw, provided that the honey is of good quality. Boiled melicrat is weaker than raw, and produces less bulky stools, neither of which qualities is asked for in melicrat. But preferably it should be employed in this (sc. boiled) state when the honey happens to be of poor quality, unclean, dark, and unpleasant smelling, since boiling will remove most of these offensive defects.

58. The drink called oxymel you will find to be useful in many ways in acute diseases. It stimulates the expectoration of sputum and eases breathing. The proper occasions for its employment are the following. Very acidic oxymel will be of no little value for sputa being coughed up only with difficulty, for if it brings up the sputum that is causing hawking, and lubricates and cleans the windpipe like a feather, it will relieve the lung with its soothing effect: if it succeeds in this way, it will render a great service. But sometimes very acidic oxymel is not able to compel the expectoration of such sputum, but even makes it more viscid and causes harm. This happens in most cases in patients who are altogether in a mortal state, and no longer have the strength to cough and expel the material being held in their lung. Thus, in this case you must reassess the person's strength, and if there seems to be hope, to give the drink. Give it, if you do, somewhat warmed, a little at a time, and not in too great a gush.

59. The slightly acidic variety (sc. of oxymel) moistens the mouth and throat, promotes the expectoration of sputum, and quenches thirst. It soothes the hypochondrium

---

63 ἥδιον A: ῥητΐδιον MV   64 ὕποξυ A: ἔποξυ MV

μενές· καὶ τὰς ἀπὸ μέλιτος βλάβας κωλύει· τὸ γὰρ ἐν μέλιτι χολῶδες κολάζεται. ἔστι δὲ καὶ φυσέων καταρρηκτικὸν καὶ ἐς οὔρησιν προτρεπτικόν· ἐντέρου μέντοι
354 τῷ κάτω μέρει πλαδαρώτερον | καὶ ξύσματα ἐμποιέει· ἔστι δ᾽ ὅτε καὶ φλαῦρον τοῦτο ἐν τῇσιν ὀξείῃσιν νούσοισι, μάλιστα μὲν ὅτι φύσας κωλύει περαιοῦσθαι, ἀλλὰ παλινδρομέειν ποιέει. ἔτι δὲ καὶ ἄλλως γυιοῖ καὶ ἀκρωτήρια ψύχει· ταύτην καὶ οἶδα μούνην βλάβην δι᾽ ὀξυμέλιτος γινομένην, ἥτις ἀξίη γραφῆς.

60. Ὀλίγον δὲ τὸ τοιόνδε ποτὸν νυκτὸς μὲν καὶ νήστει πρὸ ῥυφήματος ἐπιτήδειον προπίνεσθαι· ἀτὰρ καὶ ὁκόταν πολὺ μετὰ ῥύφημα ᾖ, οὐδὲν κωλύει πίνειν. τοῖσι δὲ ποτῷ μοῦνον διαιτωμένοισιν ἄνευ ῥυφημάτων διὰ τόδε οὐκ ἐπιτήδειόν ἐστιν αἰεὶ διὰ παντὸς
356 χρέεσθαι | τούτῳ. μάλιστα μὲν διὰ ξύσιν καὶ τρηχυσμὸν τοῦ ἐντέρου· ἀκόπρῳ γὰρ ἐόντι μᾶλλον ἐμποιοίη[65] ἂν ταῦτα κενεαγγίης παρεούσης· ἔπειτα δὲ καὶ τὸ μελίκρητον τῆς ἰσχύος ἀφαιρέοιτ᾽ ἄν. ἢν μέντοι ἀρήγειν φαίνηται πρὸς τὴν ξύμπασαν νοῦσον πολλῷ ποτῷ τούτῳ χρέεσθαι, ὀλίγον χρὴ τὸ ὄξος παραχέειν, ὅσον μοῦνον γινώσκεσθαι· οὕτω γὰρ καὶ ἃ φιλέει βλάπτειν, ἥκιστα ἂν βλάπτοι, καὶ ἃ δεῖται ὠφελείης, προσωφελοίη ἄν.

358 61. Ἐν κεφαλαίῳ δὲ | εἰρῆσθαι, αἱ ἀπὸ ὄξεος ὀξύτητες πικροχόλοισι μᾶλλον ἢ μελαγχολικοῖσι ξυμφέρουσι· τὰ μὲν γὰρ πικρὰ διαλύεται καὶ ἐκφλεγματοῦται μετεωριζόμενα ὑπ᾽ αὐτοῦ· τὰ δὲ μέλανα ζυμοῦται καὶ μετεωρίζεται καὶ πολλαπλασιοῦται·

and the viscera in that region, and prevents the ill effects of honey by checking its bilious tendencies. It also eases the passage of flatus, and is diuretic. It does, however, make the lower part of the intestine moister, and causes shreds of flesh to appear in the stools. Sometimes it may even have a bad effect in acute diseases, in particular by blocking the passage of flatus, and making it run backward. Besides, it may also weaken and cool the extremities. These are the only significant bad effects of oxymel that are worth recording.

60. It is useful to give a little of this drink (i.e., oxymel) at night when a patient is fasting before he takes gruels; also there is nothing to prevent him from drinking it long after he has taken gruel. But in patients who are being managed with drink alone and no gruels, I advise against employing oxymel continuously all through the illness, for the following reasons. Primarily because it ulcerates and roughens the intestine, having this effect particularly in a patient with no stools in his cavity because he is fasting; then it would also reduce the melicrat's strength. If, however, it seems probable that it would benefit the disease as a whole to employ a considerable amount of melicrat, reduce the amount of vinegar, so that it can just be noticed. In this way, what has the tendency to do harm will do the least harm, and what is needed to do good will do good.

61. In short, it may be said that the acidity of vinegar benefits persons with bitter bile more than those with dark bile (melancholy), since it causes the bitter humor to rise up, dissolve, and turn to phlegm, while the dark humor is fermented, raised up, and multiplied, since vinegar is an

65 *ἐμποιοίη* Kühlewein: *ἐμπνοίη* M: *ἐμπυοίη* A: *ἐμποιῆν* V

ἀναγωγὸν γὰρ μελάνων ὄξος. γυναιξὶ δὲ τὸ ἐπίπαν πολεμιώτερον ἢ ἀνδράσιν ὄξος· ὑστεραλγὲς γάρ ἐστιν.

62. (17 L.) Ὕδατι δὲ ποτῷ ἐν τῇσιν ὀξείῃσι νούσοισιν ἄλλο μὲν οὐδὲν ἔχω ἔργον ὅ τι προσθέω· οὔτε γὰρ βηχὸς παρηγορικόν ἐστιν ἐν τοῖσι περιπλευμονικοῖσιν οὔτε πτυάλου ἀναγωγόν, ἀλλ' ἧσσον τῶν ἄλλων, εἴ τις διὰ παντὸς ποτῷ ὕδατι χρέοιτο· μεσηγὺ μέντοι ὀξυμέλιτος καὶ μελικρήτου ὕδωρ ἐπιρρυφεόμενον ὀλίγον πτυάλου ἀναγωγόν ἐστι διὰ τὴν μεταβολὴν τῆς ποιότητος τῶν ποτῶν· πλημμυρίδα γάρ τινα ἐμποιέει. ἄλλως δὲ οὐδὲ δίψαν παύει, ἀλλ' ἐπιπικραίνει· χολῶδες γὰρ φύσει χολώδει καὶ ὑποχονδρίῳ κακόν. 360 κάκιστον δ' | ἑωυτοῦ καὶ χολωδέστατον καὶ φιλαδυναμώτατον, ὅταν ἐς κενεότητα ἐσέλθῃ. καὶ σπληνὸς δὲ αὐξητικὸν καὶ ἥπατός ἐστιν, ὁκόταν πεπυρωμένον ᾖ, καὶ ἐγκλυδαστικόν τε καὶ ἐπιπολαστικόν· βραδύπορον γὰρ διὰ τὸ ὑπόψυχρον εἶναι καὶ ἄπεπτον, καὶ οὔτε διαχωρητικὸν οὔτε διουρητικόν. προσβλάπτει δέ τι καὶ διὰ τόδε, ὅτι ἄκοπρόν ἐστι φύσει. ἢν δὲ δὴ καὶ ποδῶν ποτε ψυχρῶν ἐόντων ποθῇ, πάντα ταῦτα πολλαπλασίως βλάπτει, ἐς ὅ τι ἂν αὐτῶν ὁρμήσῃ.

63. Ὑποπτεύσαντι μέντοι ἐν ταύτῃσι τῇσι νούσοισι καρηβαρίην ἰσχυρὴν ἢ φρενῶν ἄψιν παντάπασιν οἴνου ἀποσχετέον. ὕδατι δ' ἐν τῷ τοιῷδε χρηστέον ἢ ὑδαρέα καὶ κιρρὸν[66] οἶνον παντελῶς δοτέον καὶ

[66] κιρρὸν A: λευκὸν MV

emetic that brings up dark humors. For women vinegar is more injurious in general than for men, since it causes pain in the uterus.

62. For water as a drink in acute diseases I have no further role to recommend. It neither soothes the coughing of pneumonias, nor brings up sputum; it has in fact less use than other drinks, if it is used as drink through the whole course of the disease. However, if it is administered between oxymel and melicrat, water has a slight role to play in promoting the expectoration of sputum, by changing the quality of the drinks by causing a kind of flooding. But otherwise it does not abate the thirst, but rather makes it keener, since it is bilious for a (sc. person with a) bilious nature, and bad for the hypochondrium. But it is worst, most productive of bile, and most debilitating when it enters (sc. a body that happens to be in) an empty state. It enlarges the spleen and liver when they are febrile, it produces splashing, and it is apt to rise to the surface. It passes out slowly, because of its coldness and lack of concoction, and it promotes movement of neither the stools nor the urines: it even interferes with them to a degree, because it makes no natural contribution to the stools. And indeed, when it is drunk while the feet are chilled, it harms many times more, in whichever of all these directions it happens to rush.

63. If a severe heaviness in the head or derangement of the mind is suspected in these diseases, wine must be completely forbidden; use water in this case, or through the whole course give a dilute tawny wine which is com-

362 ἄνοσμον | παντάπασι, καὶ μετὰ τὴν πόσιν αὐτοῦ ὕδωρ μεταποτέον ὀλίγον· ἧσσον γὰρ ἂν οὕτω τὸ ἀπὸ τοῦ οἴνου μένος ἅπτοιτο τῆς κεφαλῆς καὶ γνώμης. ἐν οἷσι δὲ μάλιστα αὐτῷ ὕδατι ποτῷ χρηστέον καὶ ὁκότε πολλῷ κάρτα καὶ ὅκου μετρίῳ, καὶ ὅπου θερμῷ καὶ ὅκου ψυχρῷ, τὰ μέν που πρόσθεν εἴρηται, τὰ δ' ἐν αὐτοῖσι τοῖσι καιροῖσι ῥηθήσεται.

64. Κατὰ ταῦτα δὲ καὶ περὶ[67] τῶν ἄλλων ποτῶν, οἷον τὸ κρίθινον καὶ τὰ ἀπὸ χλοιῆς ποιεύμενα καὶ τὰ ἀπὸ 364 σταφίδος καὶ στεμφύλων καὶ πυρῶν | καὶ κνήκου καὶ μύρτων καὶ ἀπὸ ῥοιῆς καὶ τῶν ἄλλων· ὅτε οὖν ἄν τινος αὐτῶν καιρὸς ᾖ χρέεσθαι, γεγράψεται παρ' αὐτῷ τῷ νοσήματι, ὅκωσπερ καὶ τἆλλα τῶν συνθέτων φαρμάκων.

65. (18 L.) Λουτρὸν δὲ συχνοῖσι τῶν νοσημάτων ἀρήγοι ἂν χρεωμένοισιν ἐς τὰ μὲν ξυνεχέως, ἐς τὰ δ' οὔ. ἔστι δ' ὅτε ἧσσον χρηστέον διὰ τὴν ἀπαρασκευασίην τῶν ἀνθρώπων· ἐν ὀλίγῃσι γὰρ οἰκίῃσι παρεσκεύασται τὰ ἄρμενα καὶ οἱ θεραπεύοντες ὡς δεῖ. εἰ δὲ μὴ παγκάλως λούοιτο, βλάπτοιτο ἂν οὐ σμικρά· καὶ γὰρ σκέπης ἀκάπνου δεῖ καὶ ὕδατος δαψιλέος καὶ τοῦ λουτροῦ συχνοῦ καὶ μὴ λίην λάβρου, ἢν μὴ οὕτω δέῃ. καὶ μᾶλλον μὲν μὴ σμήχεσθαι· ἢν δὲ σμήχηται, 366 θερμῷ χρέεσθαι αὐτῷ καὶ πολλαπλασίῳ ἢ ὡς | νομίζεται σμήγματι, καὶ προσκαταχεῖσθαι μὴ ὀλίγῳ, καὶ ταχέως μετακαταχεῖσθαι. δεῖ δὲ καὶ τῆς ὁδοῦ βραχείης ἐς τὴν πύαλον, καὶ ἐς εὐέμβατόν τε καὶ εὐέκβατον· εἶναι δὲ καὶ τὸν λουόμενον κόσμιον καὶ σιγηλὸν

pletely odorless, and follow the wine with a little water, since then the wine spirit will affect the head and the mind less. The most important times when drinks must be limited to water alone, when it should be given in a generous and when in a moderate amount, when drink should be warm and when cold, have all been partly explained above, and will be clarified further on the relevant occasions below.

64. Likewise, concerning other drinks such as barley water, potions made from herbs, and those prepared from raisins, pressed grapes, wheat, safflower, myrtle berries, and pomegranates etc., the correct time for the application of each of these will be recorded together with (sc. my account of) each specific disease; so also concerning the other compound drugs.

65. Bathing will be beneficial in many of the acute diseases, sometimes carried out continuously, at other times intermittently. But there are times when this is less feasible due to people's lack of facilities: for few houses have the necessary apparatus and attendants, and if the bath is not conducted properly, the harm it does will be considerable. Needed are a sheltered place free of smoke, abundant water, and a continuous flow that is not too rapid, unless such happens to be required. It is better to avoid rubbing the patient with soap, but if this is done, apply the soap warm and many times more in amount than is usual, pour on an abundant amount of additional water, and make an affusion right after that. Also, the path to the basin must be short, and the basin easy to enter and to leave. The bather should remain quiet and silent, not con-

---

[67] Κατὰ ταῦτα . . . περὶ A: Περὶ δὲ MV

καὶ μηδὲν αὐτὸν προσεξεργάζεσθαι, ἀλλ᾽ ἄλλους καὶ καταχέειν καὶ σμήχειν· καὶ μετακέρασμα πολλὸν ἡτοιμάσθαι καὶ τὰς ἐπαντλήσιας ταχείας ποιέεσθαι· καὶ σπόγγοισι χρέεσθαι ἀντὶ στεγγίδος, καὶ μὴ ἄγαν ξηρὸν χρίεσθαι τὸ σῶμα. κεφαλὴν μέντοι ἀνεξηράνθαι χρὴ ὡς οἷόν τε μάλιστα ὑπὸ σπόγγου ἐκμασσο-
368 μένην. καὶ μὴ διαψύχεσθαι τὰ ἄκρα μήτε τὴν | κεφαλὴν μήτε τὸ ἄλλο σῶμα· καὶ μήτε νεορρύφητον μήτε νεόποτον λούεσθαι μηδὲ ῥυφέειν μηδὲ πίνειν ταχὺ μετὰ τὸ λουτρόν.

66. Μέγα μὲν δὴ μέρος χρὴ νέμειν τῷ κάμνοντι, ἢν ὑγιαίνων ᾖ φιλόλουτρος ἄγαν καὶ εἰθισμένος λούεσθαι· καὶ γὰρ ποθέουσι μᾶλλον οἱ τοιοίδε καὶ ὠφελέονται λουσάμενοι καὶ βλάπτονται μὴ λουσάμενοι. ἁρμόζει δ᾽ ἐν περιπλευμονίῃσι μᾶλλον ἢ ἐν καύσοισι τὸ ἐπίπαν· καὶ γὰρ ὀδύνης τῆς κατὰ πλευρὴν καὶ στήθεος καὶ μεταφρένου παρηγορικόν ἐστι τὸ λουτρὸν καὶ πτυάλου πεπαντικὸν καὶ ἀναγωγὸν καὶ εὔπνοον καὶ ἄκοπον· μαλθακτικὸν γὰρ καὶ ἄρθρων καὶ τοῦ ἐπιπολαίου δέρματος· καὶ οὐρητικὸν δὲ καὶ καρηβαρίην λύει καὶ ῥῖνας ὑγραίνει.

67. Ἀγαθὰ μὲν οὖν λουτρῷ τοσαῦτα πάρεστιν, ὧν πάντων δεῖ. εἰ μέντοι τῆς παρασκευῆς ἔνδειά τις
370 ἔσται ἑνὸς ἢ πλειόνων, κίνδυνος μὴ λυσιτελέειν | τὸ λουτρόν, ἀλλὰ μᾶλλον βλάπτειν· ἓν γὰρ ἕκαστον αὐτῶν μεγάλην φέρει βλάβην μὴ προπαρασκευασθὲν ὑπὸ τῶν ὑπουργῶν ὡς δεῖ. ἥκιστα δὲ λούειν καιρὸς τούτους, οἷσιν ἡ κοιλίη ὑγροτέρη τοῦ καιροῦ ἐν τῇσι

tributing anything himself, but leave it to others to pour on the water and do the rubbing. A large amount of mixed cold and hot water must be prepared, and the affusions made rapidly. Sponges should be used rather than a scraper, and the body must be anointed with oil before it becomes too dry; the head, however, must be dried as thoroughly as possible, best by rubbing it with a sponge. Neither the extremities, nor the head, nor the rest of the body should be allowed to become too cold. Neither should the bath be conducted just after a patient has taken a gruel or a drink, nor should a gruel or a drink be taken just after the bath.

66. A great part of the decision (sc. about bathing) must be left to the patient, depending upon whether he is very fond of baths when he is in good health, and if he is in the habit of taking a bath. For such people long more to bathe, and benefit more from doing so, or are more harmed by not doing so. In general, the bath is more suitable in pneumonias than in ardent fevers, since it soothes the pain in the side, as well as in the chest and back; besides, it helps the sputum to be concocted and coughed up, it aids breathing, and it relieves fatigue. Further, it softens the joints and the surface of the skin, acts as a diuretic, relieves heaviness in the head, and moistens the nostrils.

67. Such then are the good points of bathing, and they are all necessary. If any of the preparations are neglected, either one or more of them, there is a danger that the bath will bring no benefit, but rather cause harm. Any of these measures will do significant harm if it is not carried out properly by the attendants in advance. It is least suitable to bathe those whose cavity is too moist in acute diseases,

νούσοισιν· ἀτὰρ οὐδ' οἷσιν ἕστηκε μᾶλλον τοῦ καιροῦ καὶ μὴ προδιελήλυθεν. οὐδὲ δὴ τοὺς γεγυιωμένους χρὴ λούειν οὐδὲ τοὺς ἀσώδεας ἢ ἐμετικοὺς οὐδὲ τοὺς ἐπανερευγομένους χολῶδες οὐδὲ τοὺς αἱμορραγέοντας ἐκ ῥινῶν, εἰ μὴ ἔλασσον τοῦ καιροῦ ῥέοι· τοὺς δὲ καιροὺς οἶδας. ἢν δὲ ἔλασσον τοῦ καιροῦ ῥέῃ, λούειν, ἤν τε ὅλον τὸ σῶμα πρὸς τὰ ἄλλα ἀρήγῃ, ἤν τε τὴν κεφαλὴν μοῦνον.

68. *Ἢν οὖν αἵ τε παρασκευαὶ ἔωσιν ἐπιτήδειοι καὶ ὁ κάμνων μέλλῃ εὖ δέξασθαι τὸ λουτρόν, λούειν χρὴ*
372 *ἑκάστης ἡμέρης· τοὺς δὲ φιλολουτρέοντας,* | *οὐδ' εἰ δὶς τῆς ἡμέρης λούοις, οὐδὲν ἂν βλάπτοις.*[68] *χρέεσθαι δὲ λουτροῖσι τοῖσιν οὔλῃσι πτισάνῃσι χρεωμένοισι*[69] *παρὰ πολὺ μᾶλλον ἐνδέχεται, ἢ τοῖσι χυλῷ μοῦνον χρεωμένοισιν· ἐνδέχεται δὲ καὶ τούτοισιν ἐνίοτε· ἥκιστα δὲ καὶ τοῖσι ποτῷ μοῦνον χρεωμένοις· ἔστι δ' οἷσι καὶ τούτων ἐνδέχεται. τεκμαίρεσθαι δὲ χρὴ τοῖσι προγεγραμμένοισιν, οὕς τε μέλλει λουτρὸν ὠφελέειν ἐν ἑκάστοισι τῶν τρόπων τῆς διαίτης οὕς τε μή· οἷσι*
374 *μὲν γὰρ προσδέεταί* | *τινος κάρτα τούτων, ὁκόσα λουτρὸν ἀγαθὰ ποιέει, λούειν καθ' ὅσα ἂν λουτρῷ ὠφε-*
376 *λέηται· οἷσι δὲ τούτων* | *μηδενὸς προσδεῖ καὶ πρόσεστιν αὐτοῖσι τῶν σημείων, ἐφ' οἷς λούεσθαι οὐ ξυμφέρει, οὐ δεῖ λούειν.*

[68] *οὐδὲν . . . βλάπτοις* A: *ούκ ἂν ἁμάρτοις* MV
[69] *τοῖσιν οὔλ. . . . χρεωμένοισι* A: om. MV

but also those whose cavity is excessively costive and has not previously had a movement. You must also refuse the bath to patients who are very weak, who are nauseated or have vomited, who have bilious eructations, or who are hemorrhaging from their nostrils, unless the amount of the flux is abnormally little—you will know how much that is. If the epistaxis is so little, bathe either the whole body, if this is indicated for other reasons, or just the head.

68. Thus, if the requisite conditions are met, and the patient is well disposed to take the bath, you should have him bathe every day; persons fond of bathing you would not harm even if it was twice a day. Those taking unstrained barley are much more able to tolerate having a bath than those taking only barley water, although it is sometimes acceptable to the latter too; it is least acceptable for people taking only drinks, although even some of these too can bathe. You must make an assessment, based on what has been said, who is likely to profit from a bath in each kind of regimen, and who not. In patients in great need of some of the advantages that the bath brings, you should allow bathing to the extent that it helps them. But to patients who have no need of such advantages, and who besides have any signs indicating that they will not profit from bathing, you should refuse it.

SACRED DISEASE

INTRODUCTION

Sacred Disease provides an account of epilepsy (and perhaps other convulsive disorders) in terms of the same natural processes that explain other diseases. In the first four chapters the author denies any special sacredness to the sacred disease,[1] by demonstrating that many diseases have features that are sacred in the same sense, and that this disease has the same kind of nature and pathophysiological cause as the rest.[2] The remaining chapters of the treatise present a detailed exposition of this nature and cause:

5 The sacred disease is inherited in families, and is more common in persons suffering from excess phlegm than in those suffering from excess bile.

6/7 The sacred disease arises from the brain and vessels of the body, in which the movement of breath produces sensation.

[1] See "sacred" in the Note on Technical Terms in the General Introduction, and "sacred disease" in Loeb *Hippocrates* vol. 1 (LCL 147), p. lxxiv.

[2] For a discussion of the polemics in this treatise, see Ducatillon, pp. 159–266.

8 The sacred disease begins in the fetus in utero, when impurities in its brain are not expelled as they should be, so that it becomes phlegmatic.

9 If later this phlegm flows down on to the heart and lung, cooling and congealing the blood in the vessels, seizures result.

10 Explanation of the various symptoms produced by this blockage.

11/12 Features of seizures in children, and in older persons.

13 Particular situations and factors that produce seizures.

14 How repeated seizures tend to become incurable.

15 The aura experienced by patients before an attack.

16 The effects of particular winds on the sacred disease.

17 The brain as the site of all mental function, and the effect of moisture on it.

18 How phlegm and bile cause different kinds of madness by impairing the brain.

19 How air acting though the brain gives rise to intelligence and volition.

20 Why the brain (not the diaphragm/heart) is the site of mental function.

21 Conclusion: the sacred disease is no more sacred than other diseases, and no less natural. Medicine is able to treat diseases without purifications and magic.

Erotian names the *Sacred Disease* in his Hippocratic census, among the "etiological and scientific works," and includes about a dozen terms from it in his *Glossary*, seven of which are definitely taken out of this treatise;[3] one of these entries (A47, ἀλάστορες) makes reference to the first *syntaxis* of Bacchius of Tanagra's *Words*.[4] Three of the glosses in Galen's Hippocratic *Glossary* are certainly derived from *Sacred Disease*, and possibly five more.[5] Otherwise, knowledge of the work does not seem to have been widespread in antiquity.[6]

The Greek text is transmitted in the two independent manuscripts M and Θ, while the fifteenth-century Corsinianus 1410 (= Co) kept in the library of the Accademia dei Lincei in Rome occupies a stemmatic position between M and Θ, apparently as the result of contamination; Co may also contain variants from some further independent source.[7] *Sacred Disease* is present in the Hippocratic collected editions and translations, and has been the subject of several special studies:

Dietz, Fr. *Hippocratis De morbo sacro*. Leipzig, 1827. (= Dietz)

Wilamowitz, U. *Griechisches Lesebuch*: Text II, pp. 269–77. Berlin, 1902.

[3] Cf. Nachmanson, pp. 325–28.

[4] Cf. Erotian, p. 17; Loeb *Hippocrates* vol. 1 (LCL 147), pp. xlvi–xlix.

[5] Cf. Perilli: α48, α78, ο24; α33, μ21, ρ20, φ24, φ25.

[6] See *Testimonien* vol. I, pp. 330–38; vol. II,1, p. 340f.; vol. III, pp. 283–85.

[7] See Rivier, pp. 19–68; Diller *KSAM*, pp. 238–46; Jouanna *Morb. sacr.*, pp. lxxxvii–xciv.

Temkin, O. *The Falling Sickness*. Baltimore, 1945; 1971[2], pp. 4–27.

Rivier, A. *Recherches sur la tradition manuscrite du traité hippocratique "De morbo sacro."* Berne, 1962. (= Rivier)

Grensemann, H. *Die hippokratische Schrift "Über die heilige Krankheit."* Berlin, 1968. (= Grensemann)

Roselli, A. *Ippocrate. La malattia sacra*. Venice, 1996.

Jouanna, J. *Hippocrate. De la maladie sacrée*. Budé II (3). Paris, 2003. (= Jouanna *Morb. sacr.*)

The present edition depends on all these works, but on Jouanna's in particular. The chapter numbering followed is that of Ermerins, Fuchs, and Jones, with Littré's (= Jouanna's) in brackets where it is different.

ΠΕΡΙ ΙΕΡΗΣ ΝΟΥΣΟΥ

VI 352 Littré

1. Περὶ τῆς ἱερῆς νούσου καλεομένης ὧδ᾽ ἔχει. οὐδέν τί μοι δοκέει τῶν ἄλλων θειοτέρη εἶναι νούσων οὐδὲ ἱερωτέρη, ἀλλὰ φύσιν μὲν ἔχει [καὶ τὰ λοιπὰ νοσήματα ὅθεν γίνεται, φύσιν δὲ αὐτὴ][1] καὶ πρόφασιν. οἱ δ᾽ ἄνθρωποι ἐνόμισαν θεῖόν τι πρῆγμα εἶναι ὑπὸ ἀπειρίης καὶ θαυμασιότητος, ὅτι οὐδὲν ἔοικεν ἑτέροισι. καὶ κατὰ μὲν τὴν ἀπορίην αὐτοῖσι τοῦ μὴ γινώσκειν τὸ θεῖον διασῴζεται, κατὰ δὲ τὴν εὐπορίην τοῦ τρόπου τῆς ἰήσιος ᾧ ἰῶνται,[2] ἀπόλλυται, ὅτι καθαρμοῖσί τε ἰῶνται καὶ ἐπαοιδῇσιν. εἰ δὲ διὰ τὸ θαυμάσιον θεῖον νομιεῖται, πολλὰ τὰ ἱερὰ νοσήματα ἔσται καὶ οὐχὶ ἕν, ὡς ἐγὼ ἀποδείξω ἕτερα οὐδὲν ἧσσον ἐόντα

354

θαυμάσια οὐδὲ τερατώδεα,[3] | ἃ οὐδεὶς νομίζει ἱερὰ εἶναι. τοῦτο μὲν οἱ πυρετοὶ οἱ ἀμφημερινοὶ καὶ οἱ τριταῖοι καὶ οἱ τεταρταῖοι οὐδὲν ἧσσόν μοι δοκέουσιν ἱεροὶ εἶναι καὶ ὑπὸ θεοῦ γίνεσθαι ταύτης τῆς νούσου, ὧν οὐ θαυμασίως ἔχουσιν· τοῦτο δὲ ὁρῶ μαινομένους ἀνθρώπους καὶ παραφρονέοντας ἀπὸ οὐδεμιῆς προφάσιος ἐμφανέος, καὶ πολλά τε καὶ ἄκαιρα ποιέοντας,

SACRED DISEASE

1. With regard to the sacred disease, as it is called, the matter is as follows. In my opinion, it is not a bit more divine or sacred than other diseases, but has both a nature and a cause. People just think it is divine due to their lack of experience and love of wonders, because it is not at all like other diseases. While their inability to understand it preserves its divinity, their simplistic way of treating it, with purifications and incantations, negates its divinity. And if this disease is held to be divine because it seems marvelous, there will be many sacred diseases, not one, for as I am about to show, there are other diseases that are no less marvelous or prodigious, although no one believes they are sacred. For instance, quotidian fevers, tertians and quartans seem to me no less sacred and the work of a god than this disease, none of which is held to be marvelous. Another example: I see people who are mad and out of their mind from no apparent cause, doing many improper things; and I know that many wail and shout in

---

[1] Del. Jones as an intruded marginal note; Grensemann deletes the whole sentence οὐδέν τί μοι . . . καὶ πρόφασιν. Cf. the beginning of ch. 5 below.

[2] ᾧ ἰῶνται Littré: ἰῶνται M: ὠπῶνται Θ

[3] οὐδὲ τερατώδεα Aldina: οὐδέτερα τῷδε M: om. Θ

ἔν τε τῷ ὕπνῳ οἶδα πολλοὺς οἰμώζοντας καὶ βοῶντας, τοὺς δὲ πνιγομένους, τοὺς δὲ καὶ ἀναΐσσοντάς τε καὶ φεύγοντας ἔξω, καὶ παραφρονέοντας μέχρι ἐπέγρωνται, ἔπειτα δὲ ὑγιέας ἐόντας καὶ φρονέοντας ὥσπερ καὶ πρότερον, ἐόντας τ' αὐτοὺς ὠχρούς τε καὶ ἀσθενέας· καὶ ταῦτα οὐχ ἅπαξ, ἀλλὰ πολλάκις. ἄλλα τε πολλά ἐστι καὶ παντοδαπὰ ὧν περὶ ἑκάστου λέγειν πολὺς ἂν εἴη λόγος.

2. Ἐμοὶ δὲ δοκέουσιν οἱ πρῶτοι τοῦτο τὸ νόσημα ἀφιερώσαντες τοιοῦτοι εἶναι ἄνθρωποι οἷοι καὶ νῦν εἰσι μάγοι τε καὶ καθάρται καὶ ἀγύρται καὶ ἀλαζόνες, οὗτοι δὲ καὶ προσποιέονται σφόδρα θεοσεβέες εἶναι καὶ πλέον τι εἰδέναι. οὗτοι τοίνυν παραμπεχόμενοι καὶ προβαλλόμενοι τὸ θεῖον τῆς ἀμηχανίης τοῦ μὴ ἔχειν ὅ τι προσενέγκαντες ὠφελήσουσι, ὡς[4] μὴ κατάδηλοι ἔωσιν οὐδὲν ἐπιστάμενοι, ἱερὸν ἐνόμισαν τοῦτο τὸ πάθος εἶναι, καὶ λόγους ἐπιλέξαντες ἐπιτηδείους τὴν ἴησιν κατεστήσαντο ἐς τὸ ἀσφαλὲς σφίσιν αὐτοῖσι, καθαρμοὺς προσφέροντες καὶ ἐπαοιδάς, λουτρῶν τε ἀπέχεσθαι κελεύοντες καὶ ἐδεσμάτων πολλῶν
356 καὶ ἀνεπιτηδείων ἀνθρώποισι νοσέουσιν | ἐσθίειν· θαλασσίων μὲν τρίγλης, μελανούρου, κεστρέος, ἐγχέλυος (οὗτοι γὰρ ἐπικηρότατοί εἰσιν[5]), κρεῶν δὲ αἰγείων[6] καὶ ἐλάφων καὶ χοιρίων καὶ κυνός (ταῦτα γὰρ κρεῶν ταρακτικώτατά ἐστι τῆς κοιλίης), ὀρνίθων δὲ ἀλεκτρυόνος καὶ τρυγόνος καὶ ὠτίδος (ἅ νομίζεται ἰσχυρότατα εἶναι), λαχάνων δὲ μίνθης, σκορόδου καὶ κρομμύων (δριμὺ γὰρ ἀσθενέοντι οὐδὲν συμφέρει),

their sleep, that others choke, and others jump up and rush out of the house, all in a state of derangement until they wake up, but then are just as healthy and rational as they were in the first place, except for being pale and weak; and these things happen not once, but in many cases. There are also many other examples of various kinds, which would each require a long account to explain.

2. I believe that the first people to regard this disease as sacred were what are still today called magicians, purifiers, mendicants, and charlatans, people who present themselves as being especially pious and knowledgeable. Being in fact helpless and having nothing useful to offer, such people cloak themselves and hide behind superstition in order to prevent their ignorance from being exposed: they deem the disease to be sacred, and, by contriving a plausible argument, set up a method of treatment that serves their own purposes: they prescribe purifications and incantations, prohibit the use of baths and many foods unsuitable for the ill to eat—e.g., of sea foods the red mullet, black-tail, gray mullet, and eel (since these are the most dangerous), of meats those of goats, deer, pork and dog (the most upsetting of meats for the cavity), of fowls the chicken, turtledove, and bustard (which are considered to be the strongest), and of plants mint, garlic, and onion (since sharp foods are of no benefit to the sick)—

---

[4] ὡ̔ς M: καὶ ὡ̔ς Θ

[5] ἐπικηρότατοί εἰσιν Θ: εἰσιν ἐπικαιρότατοι M

[6] Add. καὶ τύρου αἰγείου Θ

ἱμάτιον δὲ μέλαν μὴ ἔχειν (θανατῶδες γὰρ τὸ μέλαν), μηδὲ ἐν αἰγείῳ κατακέεσθαι δέρματι μηδὲ φορέειν, μηδὲ πόδα ἐπὶ ποδὶ ἔχειν, μηδὲ χεῖρα ἐπὶ χειρί· πάντα γὰρ ταῦτα κωλύματα εἶναι. ταῦτα δὲ τοῦ θείου εἵνεκα προστιθέασιν, ὡς πλέον τι εἰδότες, καὶ ἄλλας προφάσιας λέγοντες, ὅπως, εἰ μὲν ὑγιὴς γένοιτο, αὐτῶν ἡ δόξα εἴη καὶ ἡ δεξιότης, εἰ δὲ ἀποθάνοι, ἐν ἀσφαλεῖ καθισταῖντο αὐτῶν αἱ ἀπολογίαι καὶ ἔχοιεν πρόφασιν ὡς οὐδὲν αἴτιοί εἰσιν,[7] ἀλλ' οἱ θεοί· οὔτε γὰρ φαγέειν οὔτε πιέειν ἔδοσαν φάρμακον οὐδέν, οὔτε λουτροῖσι καθήψησαν, ὥστε δοκέειν αἴτιοι εἶναι. ἐγὼ δὲ δοκέω Λιβύων ἂν τῶν τὴν μεσόγειον οἰκεόντων οὐδέν' ἂν[8] ὑγιαίνειν, εἴ τι ἐπ' αἰγείοισι δέρμασιν ἢ κρέασιν ἦν, ὡς ἐκεῖ γε[9] οὐκ ἔχουσιν οὔτε στρῶμα οὔτε ἱμάτιον οὔτε ὑπόδημα ὅ τι μὴ αἴγειόν ἐστιν· οὐ γάρ ἐστιν | 358 ἄλλο προβάτιον οὐδὲν ἢ αἶγες καὶ βόες. εἰ δὲ ταῦτα ἐσθιόμενα καὶ προσφερόμενα τὴν νοῦσον τίκτει τε καὶ αὔξει καὶ μὴ ἐσθιόμενα ἴηται, οὐκέτι ὁ θεὸς αἴτιος ἐστίν, οὐδὲ οἱ καθαρμοὶ ὠφελέουσιν, ἀλλὰ τὰ ἐδέσματα τὰ ἰώμενά ἐστι καὶ τὰ βλάπτοντα, τοῦ δὲ θεοῦ ἀφανίζεται ἡ δύναμις.

3. Οὕτως οὖν ἔμοιγε δοκέουσιν οἵτινες τῷ τρόπῳ τούτῳ ἐγχειρέουσιν ἰῆσθαι ταῦτα τὰ νοσήματα οὔτε ἱερὰ νομίζειν εἶναι οὔτε θεῖα· ὅπου γὰρ ὑπὸ καθαρμῶν τοιούτων μετάστατα γίνεται καὶ ὑπὸ θεραπείης τοι-

[7] Add. αὐτοί M
[8] οὐδέν' ἂν Jones: οὐδένα M: οὐδὲν ἂν Θ

and forbid the wearing of a black garment (black is associated with death), or lying on or wearing a goatskin, or setting one foot over the other, or one hand over the other. All such things (sc. they say) are spells, and this they claim on the basis of the disease's divinity, of which they pretend to have superior knowledge, and they put forth other causes as well, so that if a person recovers, it adds to their reputation and apparent cleverness, but if the person dies, they have sure excuses and a reason why they are not in the least to blame, but that it is the gods' fault: after all, they have prescribed no drug to be eaten or drunk, nor have they fomented patients in baths, from which they could seem to be to blame. Indeed, it seems to me that no Libyan inhabiting the interior of that country could possibly be healthy, if health depended in any way on goatskins and goat's meat, since the Libyans have no bedding, clothing or shoes except those of goatskin, and the only animals they have to eat are goats and cattle. Furthermore, if these things when eaten and administered cause a disease to start and increase, and which when not eaten heal it, then the condition is no longer being caused by a god, nor is it purifications that are bringing help, but it is foods which are helping and harming, so that the power of any divinity disappears.

3. Thus I believe that those who undertake to heal these diseases in the way named must in fact hold them to be neither sacred nor divine. For if such purifications and treatments are able to bring about changes, why should

---

9 εἴ τι ἐπ᾽ . . . ἦν, ὡς ἐκεῖ γε Θ: ὅτι ἐπ᾽ . . . ἕνεκά γε ὡς M

ἧσδε, τί κωλύει καὶ ὑφ᾽ ἑτέρων τεχνημάτων ὁμοίων τούτοισιν ἐπιγίνεσθαί τε τοῖσιν ἀνθρώποισι καὶ προσπίπτειν, ὥστε τὸ θεῖον μηκέτι αἴτιον εἶναι, ἀλλά τι ἀνθρώπινον; ὅστις γὰρ οἷός τε περικαθαίρων ἐστὶ[10] καὶ μαγεύων ἀπάγειν τοιοῦτον πάθος, οὗτος κἂν ἐπάγοι ἕτερα τεχνησάμενος, καὶ ἐν τούτῳ τῷ λόγῳ τὸ θεῖον ἀπολύεται. τοιαῦτα λέγοντες καὶ μηχανώμενοι προσποιέονται πλέον τι εἰδέναι, καὶ ἀνθρώπους ἐξαπατῶσι προστιθέμενοι αὐτοῖς ἁγνείας τε καὶ καθάρσιας, ὅ τε πολὺς αὐτοῖς τοῦ λόγου ἐς τὸ θεῖον ἀφήκει καὶ τὸ δαιμόνιον. καίτοι ἔμοιγε οὐ περὶ εὐσεβείης τοὺς λόγους δοκέουσι ποιέεσθαι, ὡς οἴονται, ἀλλὰ περὶ δυσσεβείης[11] μᾶλλον, καὶ ὡς οἱ θεοὶ οὐκ εἰσί. τὸ δὲ εὐσεβὲς αὐτῶν καὶ τὸ θεῖον ἀσεβές ἐστι καὶ ἀνόσιον, ὡς ἐγὼ διδάξω.

4. Εἰ γὰρ σελήνην καθαιρέειν καὶ ἥλιον ἀφανίζειν καὶ χειμῶνά τε καὶ εὐδίην ποιέειν καὶ ὄμβρους καὶ αὐχμοὺς καὶ θάλασσαν ἄφορον καὶ γῆν καὶ τἆλλα | 360 τὰ τοιουτότροπα πάντα ὑποδέχονται[12] ἐπίστασθαι—εἴτε καὶ ἐκ τελετέων εἴτε καὶ ἐξ ἄλλης τινὸς γνώμης ἢ μελέτης φασὶ ταῦτα οἷόν τ᾽ εἶναι γενέσθαι—οἱ ταῦτ᾽ ἐπιτηδεύοντες δυσσεβέειν ἔμοιγε δοκέουσι καὶ θεοὺς οὔτε εἶναι νομίζειν οὔτε ἰσχύειν οὐδὲν, οὔτε εἴργεσθαι ἂν οὐδενὸς τῶν ἐσχάτων ποιέοντες, ὡς οὐ δεινοὶ αὐτοῖσίν εἰσιν. εἰ γὰρ ἄνθρωπος μαγεύων καὶ θύων σελήνην καθαιρήσει καὶ ἥλιον ἀφανιεῖ καὶ χειμῶνα καὶ εὐδίην ποιήσει, οὐκ ἂν ἔγωγ᾽ ἔτι[13] θεῖον νομίσαιμι τούτων εἶναι οὐδέν, ἀλλ᾽ ἀνθρώπινον, εἰ δὴ τοῦ θείου

not other similar techniques affect people and make them sick, so that the cause of their disease would no longer be divine, but something human? Indeed, a person who is capable of removing such a disease by ritual purifications and magic could surely also bring another one on by using other techniques; which again dismisses the divine. By arguing and contriving in these ways, such folk pretend to know more than they do, and they deceive people by prescribing for them ceremonies and rites, with a good part of their talk being devoted to gods and the divine. And in fact, I believe that their speeches are not a proof of piety, as they pretend, but rather of impiety, and a denial of the gods' existence. Their supposed piety and divine talk is in fact impious and a sacrilege, as I shall demonstrate.

4. For if they claim to know how to make the moon descend and the sun disappear (sc. in an eclipse), and create storms or fair weather, and rains or droughts, and make the sea and earth barren, and do all kinds of other things like this—whether they claim these things can be achieved with rites or some other knowledge or practice—people practicing such things seem to me to be impious and not to believe that the gods exist or have any power, nor would they hesitate to take any extreme actions, since they do not fear the gods. For if a person used magic and sacrifices to make the moon descend and the sun disappear, and to create storms or fair weather, I would no longer think any of these things were divine, but human, since

---

[10] ἐστὶ Θ: om. M [11] δυσ- M: ἀ- Θ

[12] ὑποδέχ- Θ: ἐπιδέχ- M

[13] ἔγωγ' ἔτι Wilamowitz: ἔγωγέ τι codd.

ἡ δύναμις ὑπὸ ἀνθρώπου γνώμης κρατέεται καὶ δεδούλωται. ἴσως δὲ οὐχ οὕτως ἔχει ταῦτα, ἀλλ' ἄνθρωποι βίου δεόμενοι πολλὰ καὶ παντοῖα τεχνῶνται καὶ ποικίλλουσιν ἔς τε τἆλλα πάντα καὶ ἐς τὴν νοῦσον ταύτην, ἑκάστῳ εἴδει τοῦ πάθεος θεῷ τὴν αἰτίην προστιθέντες. οὐ γὰρ ἐναλλά<ξ, ἀλλὰ>[14] πλεονάκις γε μὴν ταὐτὰ μεμίμηνται· ἢν μὲν γὰρ αἶγα μιμῶνται,[15] καὶ ἢν βρύχωνται, κἢν τὰ δεξιὰ σπῶνται, Μητέρα θεῶν φασιν αἰτίην εἶναι. ἢν δὲ ὀξύτερον καὶ ἐντονώτερον φθέγγηται, ἵππῳ εἰκάζουσι, καί φασι Ποσειδῶνα αἴτιον εἶναι. ἢν δὲ καὶ τῆς κόπρου τι παρίῃ, ὅσα πολλάκις γίνεται ὑπὸ τῆς νούσου βιαζομένοισιν, | Ἐνοδίης θεοῦ[16] πρόσκειται ἡ ἐπωνυμίη· ἢν δὲ πυκνότερον καὶ λεπτότερον, οἷον ὄρνιθες, Ἀπόλλων νόμιος. ἢν δὲ ἀφρὸν ἐκ τοῦ στόματος ἀφίῃ καὶ τοῖσι ποσὶ λακτίζῃ, Ἄρης τὴν αἰτίην ἔχει. οἷσι δὲ νυκτὸς δείματα παρίσταται καὶ φόβοι καὶ παράνοιαι καὶ ἀναπηδήσιες ἐκ τῆς κλίνης καὶ φεύξιες ἔξω, Ἑκάτης φασὶν εἶναι ἐπιβολὰς καὶ ἡρώων ἐφόδους. καθαρμοῖσί τε χρέονται καὶ ἐπαοιδῇσι, καὶ ἀνοσιώτατόν τε καὶ ἀθεώτατον πρῆγμα[17] ποιέουσιν, ὡς ἔμοιγε δοκέει· καθαίρουσι γὰρ τοὺς ἐχομένους τῇ νούσῳ αἵματί τε καὶ ἄλλοισι τοιούτοισιν ὥσπερ μίασμά τι ἔχοντας, ἢ ἀλάστορας, ἢ πεφαρμακευμένους ὑπ' ἀνθρώπων, ἤ τι

362

[14] Conj. Jouanna *Morb. sacr.* (cf. p. 8, n. 1) [15] ἢν μὲν γὰρ αἶγα μιμῶνται om. Θ [16] Ἐνοδίης θεοῦ Grensemann: Ἐνοδείης οὐ M: ἐνοδίηι Θ [17] πρῆγμα Θ : om. M

the power of the divine had been abrogated by human intelligence and enslaved. But maybe the situation is not so, and just as people in need of a livelihood have invented and developed many and various theories in every other field, so too for this disease they have attributed the cause of each form of the condition to some god—for it is not one thing at one time and another thing at another time that diseases imitate, but often the same things. For example, if patients imitate the bleating of a goat or the roar (sc. of a lion), or have spasms on their right side, they say the Mother of the Gods is to blame; or if a patient gives out a very high, piercing cry, they liken him to a horse, and say Poseidon is to blame. If feces are passed, which often happens in people in the throes of this disease, they give it the name of the goddess Enodia.[1] But if they pass stools that are more frequent and thinner like birds', it is Apollo god of shepherds. Or if a patient expels froth from his mouth and kicks with his legs, Ares is to blame. And in patients visited at night by terrors and frights, together with derangement of the mind, leaping up out of bed, and fleeing outside, they say these are the attacks of Hecate and onslaughts of the heroes. Furthermore, such healers perform purifications and incantations involving most unholy and godless practices, in my opinion. For they purify victims of the disease with blood and such things, as if they had some defilement or were being pursued by avenging spirits, or had been bewitched by people, or had commit-

[1] Enodia: "epithet of divinities, who have their statues *by the wayside* or *at the cross-roads*, most frequently of Hecate" (LSJ).

ἔργον ἀνόσιον εἰργασμένους, οὓς ἐχρῆν τἀντία τούτων ποιέειν, θύειν τε καὶ εὔχεσθαι καὶ ἐς τὰ ἱερὰ φέροντας ἱκετεύειν τοὺς θεούς· νῦν δὲ τούτων μὲν ποιέουσιν οὐδέν, καθαίρουσι δέ, καὶ τὰ μὲν τῶν καθαρμῶν γῇ κρύπτουσι, τὰ δὲ ἐς θάλασσαν ἐμβάλλουσι, τὰ δ᾽ ἐς τὰ οὔρεα ἀποφέρουσιν, ὅπῃ μηδεὶς ἅψεται μηδὲ ἐμβήσεται· τάδ᾽ ἐχρῆν ἐς τὰ ἱερὰ φέροντας τῷ θεῷ ἀποδοῦναι, εἰ δὴ ὁ θεός ἐστιν αἴτιος. οὐ μέντοι ἔγωγε ἀξιῶ ὑπὸ θεοῦ ἀνθρώπου σῶμα μιαίνεσθαι, τὸ ἐπικηρότατον ὑπὸ τοῦ ἁγνοτάτου· | ἀλλὰ καὶ ἤν τυγχάνῃ ὑφ᾽ ἑτέρου μεμιασμένον ἤ τι πεπονθός, ὑπὸ τοῦ θεοῦ καθαίρεσθαι ἂν αὐτὸ καὶ ἁγνίζεσθαι μᾶλλον ἢ μιαίνεσθαι. τὰ γοῦν μέγιστα τῶν ἁμαρτημάτων καὶ ἀνοσιώτατα τὸ θεῖόν ἐστι τὸ καθαῖρον καὶ ἁγνίζον καὶ ἔρυμα[18] γινόμενον ἡμῖν. αὐτοί τε ὅρους τοῖσι θεοῖσι τῶν ἱερῶν καὶ τῶν τεμενέων ἀποδείκνυμεν, ὡς[19] ἂν μηδεὶς ὑπερβαίνῃ ἢν μὴ ἁγνεύῃ, ἐσιόντες τε ἡμεῖς περιρραινόμεθα οὐχ ὡς μιαινόμενοι, ἀλλ᾽ εἴ τι καὶ πρότερον ἔχομεν μύσος, τοῦτο ἀφαγνιούμενοι.[20] καὶ περὶ μὲν τῶν καθαρμῶν οὕτω μοι δοκέει ἔχειν.

364

5. (2 L.) Τὸ δὲ νόσημα τοῦτο οὐδέν τί μοι δοκέει θειότερον εἶναι τῶν λοιπῶν, ἀλλὰ φύσιν μὲν ἔχειν[21] ἣν καὶ τὰ ἄλλα νοσήματα,[22] ὅθεν ἕκαστα γίνεται, φύσιν δὲ τοῦτο καὶ πρόφασιν, καὶ ἀπὸ ταὐτοῦ θεῖον γί-

[18] ἔρυμα M: ῥύμμα Θ [19] ἀποδείκνυμεν ὡς Cornarius in marg. Ermerins: ἀποδείκνυμενοι ὡς M: δείκνυνται Θ

ted some unholy deed, when in fact they should apply treatments opposite to these, *viz.* treat them with sacrifices and prayers, and, bringing them to sanctuaries, make supplications for them to the gods. But, as it is, they do nothing like this, but instead clean them, sometimes hiding the cleanings in the earth, at other times casting them into the sea, or carrying them off to some mountain waste, where no one will touch them or walk on them. But they should by rights have carried the cleanings to sanctuaries and offered them to the god, if that god is truly the cause. I personally am convinced that a person's body could never be defiled by a god, the most corruptible thing by the most divine, and even if it happens to have been defiled by something else or injured in some way, a god would cleanse it and sanctify it rather than defiling it. For we are cleansed, blessed and protected from our most serious errors and unholy acts by the divine. And we ourselves mark the boundaries of the gods' sanctuaries and precincts to prevent anyone who is not pure from transgressing them, and when we enter them we purify ourselves all over not with the idea we are being polluted, but in case we are carrying some earlier defilement, to wash it away. This is how it stands, in my opinion, with purifications.

5. This disease is no more divine, in my opinion, than the rest, but just as each of the other diseases has a nature from which it arises, this one too results from a natural cause; furthermore, it derives its divinity from the same

---

[20] *ἀλλ᾽ εἴ τι . . . ἀφαγνιούμενοι* M: om. Θ
[21] *ἔχειν* Linden: *ἔχει* MΘ [22] *καὶ πρόφασιν* add. M

νεσθαι ἀφ᾽ ὅτευ καὶ τἆλλα πάντα, καὶ ἰητὸν εἶναι, καὶ οὐδὲν ἧσσον ἑτέρων, ὅ τι ἂν μὴ ἤδη ὑπὸ χρόνου πολλοῦ καταβεβιασμένον ᾖ, ὥστε ἤδη ἰσχυρότερον εἶναι τῶν φαρμάκων τῶν προσφερομένων. ἄρχεται δὲ ὥσπερ καὶ τἆλλα νοσήματα κατὰ γένος· εἰ γὰρ ἐκ φλεγματώδεος φλεγματώδης, καὶ ἐκ χολώδεος χολώδης γίνεται, καὶ ἐκ φθινώδεος φθινώδης, καὶ ἐκ σπληνώδεος σπληνώδης, τί κωλύει ὅτῳ πατὴρ ἢ μήτηρ εἴχετο νοσήματι, τούτῳ καὶ τῶν ἐκγόνων ἔχεσθαί τινα; ὡς ὁ γόνος ἔρχεται πάντοθεν τοῦ σώματος, ἀπό τε τῶν ὑγιηρῶν ὑγιηρός, καὶ ἀπὸ τῶν νοσερῶν νοσερός. ἕτερον δὲ μέγα τεκμήριον ὅτι οὐδὲν θειότερόν
366 ἐστι τῶν λοιπῶν νοσημάτων· | τοῖσι γὰρ φλεγματώδεσι φύσει γίνεται· τοῖσι δὲ χολώδεσιν οὐ προσπίπτει· καίτοι εἰ[23] θειότερόν ἐστι τῶν ἄλλων, τοῖσιν ἅπασιν ὁμοίως ἔδει γίνεσθαι τὴν νοῦσον ταύτην, καὶ μὴ διακρίνειν μήτε χολώδεα μήτε φλεγματώδεα.

6. (3 L.) Ἀλλὰ γὰρ αἴτιος ὁ ἐγκέφαλος τούτου τοῦ πάθεος, ὥσπερ τῶν ἄλλων νοσημάτων τῶν μεγίστων· ὅτῳ δὲ τρόπῳ καὶ ἐξ οἵης προφάσιος γίνεται, ἐγὼ φράσω σάφα. ὁ ἐγκέφαλός ἐστι τοῦ ἀνθρώπου διπλόος ὥσπερ καὶ τοῖσιν ἄλλοισι ζῴοις ἅπασιν· τὸ δὲ μέσον αὐτοῦ διείργει μῆνιγξ λεπτή· διότι οὐκ ἀεὶ κατὰ τωὐτὸ τῆς κεφαλῆς ἀλγέει, ἀλλ᾽ ἐν μέρει ἑκάτερον, ὁτὲ δὲ ἅπασαν. καὶ φλέβες δ᾽ ἐς αὐτὸν τείνουσιν ἐξ ἅπαντος τοῦ σώματος, πολλαὶ καὶ λεπταί, δύο δὲ παχεῖαι, ἡ μὲν ἀπὸ τοῦ ἥπατος, ἡ δὲ ἀπὸ τοῦ σπληνός. καὶ ἡ μὲν ἀπὸ τοῦ ἥπατος ὧδ᾽ ἔχει· τὸ μέν τι τῆς

things as all the others, and it is no less curable than the others, as long as it has not already gained strength over a long period, so as to be already stronger than the remedies that are applied. Its origin, as with other diseases, lies in heredity, for if a phlegmatic (sc. child) comes from a phlegmatic (sc. parent), a bilious one from a bilious, a consumptive from a consumptive, a splenic from a splenic, what is to prevent, when a father or a mother has any disease, one of their offspring from having the same one? After all, the seed comes from every part of a (parent's) body, healthy seed from healthy parts, diseased seed from diseased parts. Another important piece of evidence that this disease is no more divine than other diseases: it occurs naturally in phlegmatics, but does not attack the bilious: but if it is more divine than the others, such a disease should arise equally in all people, without discriminating between (sc. the two humors) bile and phlegm.

6. But in truth it is the brain which causes this affection, just as the other most serious maladies. How and why this comes about I shall now clearly describe. The human brain is double, as is that of all other living beings, and is divided in the center by a thin membrane: this is why headaches are not always on the same side, but may occur alternately on each side, and sometimes in the whole head. In the brain there terminate many fine vessels coming from the whole body, and two wide vessels, one from the liver and the other from the spleen. The vessel from the liver is as follows: one part of the vessel passes down

---

23 *καίτοι εἰ* Laurentianus Gr. 74,1 (XV c.): *καὶ τοῖσι* MΘ

φλεβὸς κάτω τείνει διὰ τῶν ἐπὶ δεξιὰ παρ' αὐτὸν τὸν νεφρὸν καὶ τὴν ψύην ἐς τὸ ἐντὸς τοῦ μηροῦ, καὶ καθήκει ἐς τὸν πόδα, καὶ καλέεται κοίλη φλέψ· ἡ δὲ ἑτέρη ἄνω τείνει διὰ φρενῶν[24] καὶ τοῦ πλεύμονος τῶν δεξιῶν· ἀπέσχισται δὲ καὶ ἐς τὴν καρδίην καὶ ἐς τὸν βραχίονα τὸν δεξιόν· καὶ τὸ λοιπὸν ἄνω φέρει διὰ τῆς κληῖδος ἐς τὰ δεξιὰ τοῦ αὐχένος, ἐς αὐτὸ τὸ δέρμα, ὥστε κατάδηλος εἶναι· παρ' αὐτὸ δὲ τὸ οὖς κρύπτεται καὶ ἐνταῦθα σχίζεται, καὶ τὸ μὲν παχύτατον καὶ μέγιστον καὶ κοιλότατον ἐς τὸν ἐγκέφαλον τελευτᾷ, τὸ δὲ ἐς τὸ οὖς τὸ δεξιόν,[25] τὸ δὲ ἐς τὸν ὀφθαλμὸν τὸν δεξιόν, τὸ δὲ ἐς τὸν μυκτῆρα. ἀπὸ μὲν τοῦ ἥπατος οὕτως ἔχει τὰ τῶν φλεβῶν. διατέταται δὲ καὶ ἀπὸ τοῦ σπληνὸς φλὲψ ἐς τὰ ἀριστερὰ καὶ κάτω καὶ ἄνω, ὥσπερ καὶ ἡ ἀπὸ τοῦ ἥπατος, λεπτοτέρη δὲ καὶ ἀσθενεστέρη. |

368 7. (4 L.) Κατὰ ταύτας δὲ τὰς φλέβας καὶ ἐσαγόμεθα[26] τὸ πολὺ τοῦ πνεύματος· αὗται γὰρ ἡμῖν εἰσιν ἀναπνοαὶ τοῦ σώματος[27] τὸν ἠέρα ἐς σφᾶς ἕλκουσαι, καὶ ἐς τὸ σῶμα τὸ λοιπὸν ὀχετεύουσι κατὰ τὰ φλέβια, καὶ ἀναψύχουσι καὶ πάλιν ἀφιᾶσιν· οὐ γὰρ οἷόν τε τὸ πνεῦμα στῆναι, ἀλλὰ ἀναχωρέει ἄνω τε καὶ κάτω. ἢν γὰρ στῇ που καὶ ἀποληφθῇ, ἀκρατὲς γίνεται ἐκεῖνο τὸ μέρος καθ' ὃ ἂν στῇ· τεκμήριον δέ· ὅταν καθημένῳ ἢ κατακειμένῳ φλέβια πιεσθῇ, ὥστε τὸ πνεῦμα μὴ διεξιέναι διὰ τῆς φλεβός, εὐθὺς νάρκη ἔχει. περὶ μὲν τῶν φλεβῶν οὕτως ἔχει.

through the right parts past that kidney and loin into the interior of the thigh, and descends to the foot—it has the name hollow vessel (*vena cava*)—while the other part runs upward through the diaphragm and lung on the right side, sending off branches into the heart and the right arm, continues up past the collarbone into the right side of the neck, rising right up to the skin so that it is visible, and then disappears beside the ear, and at that point it divides: the widest, largest, and hollowest branch terminates in the brain, while other branches go to the right ear, the right eye, and the nostril. This is the course of the vessel arising from the liver. Another vessel arises from the spleen, passes to the left side both downward and upward, like the one from the liver, but is thinner and weaker.

7. Through these vessels we take in most of our breath, as they are the air ducts of the body which draw air into themselves and conduct it to the rest of the body through small vessels, refresh it, and then expel it, since breath cannot stand still, but runs upward and downward. If breath does stand still in some part and is intercepted, the part where it is standing becomes paralyzed: proof of this is that when a person by sitting or lying compresses small vessels so that breath cannot pass through a vessel, numbness immediately follows. This is the arrangement of the vessels.

---

24 φρενῶν Dietz: φλεβῶν MΘ
25 φλέβιον λεπτὸν add. M
26 ἐσαγ. M: ἐπαγ. Θ
27 τοῦ σώματος M: ἐκ τοῦ στόματος καὶ Θ

8. (5 L.) Ἡ δὲ νοῦσος αὕτη γίνεται τοῖσι μὲν φλεγματίῃσι, τοῖσι δὲ χολώδεσιν οὔ. ἄρχεται δὲ φύεσθαι ἐπὶ τοῦ ἐμβρύου ἔτι[28] ἐν τῇ μήτρῃ ἐόντος· καθαίρεται γὰρ καὶ ἀνθέει, ὥσπερ τἆλλα μέλεα, πρὶν ἢ γενέσθαι, καὶ ὁ ἐγκέφαλος. ἐν ταύτῃ δὲ τῇ καθάρσει ἢν μὲν καλῶς καὶ μετρίως καθαρθῇ καὶ μήτε πλέον μήτ' ἔλασσον τοῦ δέοντος ἀπορρυῇ, οὕτως ὑγιηροτάτην τὴν κεφαλὴν ἔχει· ἢν δὲ πλέονα ῥυῇ ἀπὸ παντὸς τοῦ ἐγκεφάλου καὶ ἀπότηξις πολλὴ γένηται, νοσώδεά τε τὴν κεφαλὴν ἕξει αὐξόμενος καὶ ἤχου πλέην, καὶ οὔτε ἥλιον οὔτε ψῦχος ἀνέξεται· ἢν δὲ ἀπὸ ἑνός τινος γένηται ἢ ὀφθαλμοῦ ἢ ὠτός, ἢ φλέψ τις συνισχνανθῇ, 370 κεῖνο κακοῦται τὸ μέρος, ὅπως ἂν καὶ | τῆς ἀποτήξιος ἔχῃ. ἢν δὲ κάθαρσις μὴ ἐπιγένηται, ἀλλὰ ξυστραφῇ τῷ ἐγκεφάλῳ, οὕτως ἀνάγκη φλεγματώδεα εἶναι. καὶ οἷσι μὲν παιδίοις ἐοῦσιν ἐξανθέει ἕλκεα καὶ ἐς τὴν κεφαλὴν καὶ ἐς τὰ ὦτα καὶ ἐς τὸν ἄλλον χρῶτα, καὶ σιαλώδεα γίνεται καὶ μυξόρροα, ταῦτα μὲν ῥήϊστα διάγει προϊούσης τῆς ἡλικίης· ἐνταῦθα γὰρ ἀφιεῖ καὶ ἐκκαθαίρεται τὸ φλέγμα, ὃ ἐχρῆν ἐν τῇ μήτρῃ καθαρθῆναι· καὶ τὰ οὕτω καθαρθέντα[29] οὐ γίνεται ἐπίληπτα τῇ νούσῳ ταύτῃ ἐπὶ τὸ πολύ. ὅσα δὲ καθαρά τέ ἐστι, καὶ μήθ' ἕλκος μηδὲν μήτε μύξα μήτε σίαλον αὐτοῖς προέρχεται, μήτε ἐν τῇσι μήτρῃσι πεποίηται τὴν κάθαρσιν, τούτοισι δὲ ἐπικίνδυνόν ἐστιν ἁλίσκεσθαι ὑπὸ ταύτης τῆς νούσου.

9. (6 L.) Ἢν δ' ἐπὶ τὴν καρδίην ποιήσηται ὁ κατάρροος τὴν πορείην, παλμὸς ἐπιλαμβάνει καὶ ἆσθμα,

8. The sacred disease occurs in phlegmatic but not in bilious subjects. Its first growth begins in the embryo still present in the uterus, for the brain is purged there and expels its impurities before birth, as do the other parts. If this cleaning takes place successfully and in moderation, so that neither too much nor too little is expelled, a child's head will be of the healthiest order; but if more flows from the whole brain than should, and there is excessive melting, such a child's head will be unhealthy as he grows, and filled with echoing, and he will tolerate neither sun nor cold. If there is a flux from some single part such as an eye or an ear, or some vessel withers, the part in question will deteriorate in proportion to the extent of the melting. But if no cleaning takes place, but humor collects in the brain, such a person will certainly be phlegmatic. Those who as children develop lesions in their head and ears, and on the rest of their skin, and have fluxes of saliva and mucus, will go through life more easily as they grow up, for in them phlegm that should have been cleaned out while they were still in the uterus is being expelled and purged: children cleaned in this way are in most cases not subject to the sacred disease. But those who are unblemished and have no lesion, who do not expel mucus or saliva, and who were also not cleaned in the uterus, are in danger of contracting the disease.

9. If a flux makes its path to the heart, trembling begins together with difficult breathing, and the chest is disabled;

---

28 ἔτι recc.: om. MΘ

29 καθαρθέντα M: παιδευθέντα Θ

καὶ τὰ στήθεα διαφθείρεται, ἔνιοι δὲ καὶ κυφοὶ γίνονται. ὅταν γὰρ ἐπικατέλθῃ τὸ φλέγμα ψυχρὸν ἐπὶ τὸν πλεύμονα καὶ τὴν καρδίην, ἀποψύχεται τὸ αἷμα· αἱ δὲ φλέβες πρὸς βίην ψυχόμεναι πρὸς τῷ πλεύμονι καὶ τῇ καρδίῃ πηδῶσι, καὶ ἡ καρδίη πάλλεται, ὥστε ὑπὸ τῆς ἀνάγκης ταύτης τὸ ἆσθμα ἐπιπίπτειν καὶ τὴν ὀρθοπνοίην—οὐ γὰρ δέχεται τὸ πνεῦμα ὅσον[30] ἐθέλει—ἄχρι κρατηθῇ τοῦ φλέγματος τὸ ἐπιρρυὲν καὶ διαθερμανθὲν διαχυθῇ ἐς τὰς φλέβας· ἔπειτα παύεται τοῦ παλμοῦ καὶ τοῦ ἄσθματος· παύεται δὲ ὅπως ἂν καὶ τοῦ πλήθεος ἔχῃ· ἢν μὲν γὰρ πλέον ἐπικαταρρυῇ, σχολαίτερον, ἢν δ' ἔλασσον, θᾶσσον· καὶ ἢν | πυκνότεροι ἔωσιν οἱ κατάρροοι, πυκνότερα ἐπίληπτος γίνεται. ταῦτα μὲν οὖν πάσχει, ἢν ἐπὶ τὸν πλεύμονα καὶ τὴν καρδίην ἴῃ· ἢν δὲ ἐς τὴν κοιλίην, διάρροιαι λαμβάνουσιν.

372

10. (7 L.) Ἢν δὲ τούτων μὲν τῶν ὁδῶν ἀποκλεισθῇ, ἐς δὲ τὰς φλέβας, ἃς προείρηκα, τὸν κατάρροον ποιήσηται, ἄφωνος γίνεται καὶ πνίγεται, καὶ ἀφρὸς ἐκ τοῦ στόματος ἐκρέει, καὶ οἱ ὀδόντες συνηρείκασι, καὶ αἱ χεῖρες συσπῶνται, καὶ τὰ ὄμματα διαστρέφονται, καὶ οὐδὲν φρονέουσιν, ἐνίοισι δὲ καὶ ὑποχωρέει ἡ κόπρος[31] κάτω. καὶ ταῦτα γίνεται ἐνίοτε μὲν ἐς τὰ ἀριστερά, ὅτε δὲ ἐς τὰ δεξιά, ὅτε δ' ἐς ἀμφότερα.[32] ὅπως δὲ τούτων ἕκαστον πάσχει ἐγὼ φράσω· ἄφωνος μέν ἐστιν ὅταν ἐξαίφνης τὸ φλέγμα ἐπικατελθὸν ἐς[33] τὰς φλέβας ἀποκλείσῃ τὸν ἠέρα καὶ μὴ παραδέχηται

some patients also become humpbacked. For when the cold phlegm runs down on to the lung and heart, the blood is chilled. The vessels, being violently cooled, pound against the lung and the heart, and the heart palpitates, which must lead to difficult breathing and orthopnea—that is, the patient does not receive as much breath as he wants—which last until the influx of phlegm is overcome, warmed, and dispersed among the vessels. Then the palpitation and difficulty of breathing cease, stopping according to the amount of the flux: if what flowed down was greater, the recovery is more extended in time, while if it was less, the recovery is quicker. If the fluxes occur more frequently, the attacks will be more frequent. This is what happens when the flux is to the lung and heart; if it is to the cavity, diarrhea results.

10. But if the phlegm is shut out from these pathways (sc. to the lung and heart) and flows down into the vessels I have referred to, such a person becomes speechless, chokes, and foams at the mouth; his teeth are clenched, his arms are retracted, and his eyes are distorted; these patients are out of their senses, and in some, stools are even passed. Sometimes this happens on the left side of the body, sometimes on the right, and sometimes on both. How each symptom in these conditions arises I will explain. A person becomes speechless when phlegm suddenly runs into the vessels and blocks the intake of air, so

---

[30] δέχεται . . . ὅσον Θ: δέχεσθαι τὸ πνεῦμα M

[31] ἡ κόπρος M: om. Θ [32] Jones deletes καὶ ταῦτα γίνεται . . . ἀμφότερα as a note, but see Jouanna *Morb. sacr.*, p. 15, n. 1. [33] ἐς M: ἐπὶ Θ

μήτ' ἐς τὸν ἐγκέφαλον μήτ' ἐς τὰς φλέβας τὰς κοίλας μήτε ἐς τὰς κοιλίας,[34] ἀλλ' ἐπιλάβῃ τὴν ἀναπνοήν· ὅταν γὰρ λάβῃ ἄνθρωπος κατὰ τὸ στόμα καὶ τοὺς μυκτῆρας τὸ πνεῦμα, πρῶτον μὲν ἐς τὸν ἐγκέφαλον ἔρχεται, ἔπειτα δ' ἐς τὴν κοιλίην τὸ πλεῖστον μέρος, τὸ δ' ἐπὶ τὸν πλεύμονα, τὸ δ' ἐπὶ τὰς φλέβας. ἐκ τούτων δὲ σκίδναται ἐς[35] τὰ λοιπὰ μέρεα κατὰ τὰς φλέβας· καὶ ὅσον μὲν ἐς τὴν κοιλίην ἔρχεται, τοῦτο μὲν τὴν κοιλίην διαψύχει, καὶ ἄλλο οὐδὲν συμβάλλεται. ὁ δ' ἐς τὸν πλεύμονά τε καὶ ἐς τὰς φλέβας ἀὴρ ξυμβάλλεται ἐς τὰς κοιλίας ἐσιὼν [καὶ ἐς τὸν ἐγκέφαλον ἔρχεται],[36] καὶ οὕτω τὴν φρόνησιν καὶ τὴν κίνησιν τοῖσι μέλεσι παρέχει, ὥστ' ἐπειδὰν ἀποκλεισθῶσιν αἱ φλέβες τοῦ ἠέρος ὑπὸ τοῦ φλέγματος | καὶ μὴ παραδέχωνται, ἄφωνον καθιστᾶσι καὶ ἄφρονα τὸν ἄνθρωπον. αἱ δὲ χεῖρες ἀκρατέες γίνονται καὶ σπῶνται, τοῦ αἵματος ἀτρεμίσαντος καὶ οὐ διαχεομένου ὥσπερ εἰώθει. καὶ οἱ ὀφθαλμοὶ διαστρέφονται, τῶν φλεβίων ἀποκλειομένων τοῦ ἠέρος καὶ σφυζόντων. ἀφρὸς δὲ ἐκ τοῦ στόματος προέρχεται ἐκ τοῦ πλεύμονος· ὅταν γὰρ τὸ πνεῦμα μὴ ἐσίῃ ἐς αὐτόν, ἀφρίει καὶ ἀναβλύει ὥσπερ ἀποθνήσκων. ἡ δὲ κόπρος ὑπέρχεται ὑπὸ βίης πνιγομένου· πνίγεται δὲ τοῦ ἥπατος καὶ τῆς ἄνω κοιλίης πρὸς τὰς φρένας προσπεπτωκότων καὶ τοῦ στομάχου τῆς γαστρὸς ἀπειλημμένου· προσπίπτει δ' ὅταν τὸ πνεῦμα μὴ ἐσίῃ ἐς τὸ στόμα[37] ὅσον εἰώθει. λακτίζει δὲ τοῖσι ποσὶν ὅταν ὁ ἀὴρ ἀποκλεισθῇ ἐν τοῖσι σκέλεσι καὶ μὴ οἷός τε ᾖ διεκδῦναι ἔξω

374

that air is received neither into the brain, nor into the hollow vessels, nor into the bodily cavities, and inspiration is cut off. For when a person draws breath through his mouth and nostrils, it passes first to his brain, then the greatest part of it moves to his cavity, some to the lung, and some to the vessels; thence it spreads out to the remaining parts of the body through the vessels. The portion that enters the cavity cools it but makes no other contribution, but the air that goes into the lung and the vessels makes its contribution on arriving in the cavities, and in this way provides intelligence and movement to the limbs: thus, whenever these vessels are cut off from the air by phlegm, and cannot receive it, this renders the patient speechless and senseless. The arms are paralyzed and retracted when the blood is motionless and does not flow through (sc. the vessels) as usual. The eyes are distorted when their small vessels are cut off from the air, and throb. Foam coming from the lungs issues from the mouth, since when the lungs have no air they foam and bubble up as in a person who is dying. Stool descends due to the force of a person's choking, which is caused by the liver and the upper cavity falling against the diaphragm, and the mouth of the stomach being obstructed: this compression arises when breath does not enter the mouth as usual. The person kicks with his feet when air is confined in his legs and unable to escape because of the phlegm, since as it rushes

---

34 μήτε ἐς τὰς κοιλίας M: om. Θ

35 ἐς M: ἐπὶ Θ

36 καὶ . . . ἔρχεται Θ: καὶ ἐς τὸν ἐγκέφαλον M: del. Diller *Schr.*

37 στόμα M: σῶμα Θ

ὑπὸ τοῦ φλέγματος· ἀΐσσων δὲ διὰ τοῦ αἵματος ἄνω καὶ κάτω σπασμὸν ἐμποιέει καὶ ὀδύνην, διὸ λακτίζει. ταῦτα δὲ πάσχει πάντα, ὁκόταν τὸ φλέγμα παραρρυῇ ψυχρὸν ἐς τὸ αἷμα θερμὸν ἐόν· ἀποψύχει γὰρ καὶ ἵστησι τὸ αἷμα· καὶ ἢν μὲν πολὺ ᾖ τὸ ῥεῦμα καὶ παχύ, αὐτίκα ἀποκτείνει· κρατέει γὰρ τοῦ αἵματος τῷ ψυχρῷ καὶ πήγνυσιν· ἢν δ' ἔλασσον ᾖ, τὸ μὲν παραυτίκα κρατέει ἀποφράξαν τὴν ἀναπνοήν· ἔπειτα τῷ χρόνῳ ὁπόταν σκεδασθῇ κατὰ τὰς φλέβας καὶ μιγῇ τῷ αἵματι πολλῷ ἐόντι καὶ θερμῷ, ἢν κρατηθῇ οὕτως, ἐδέξαντο τὸν ἠέρα αἱ φλέβες, καὶ ἐφρόνησαν.

11. (8 L.) Καὶ ὅσα μὲν σμικρὰ παιδία κατάληπτα γίνεται τῇ νούσῳ ταύτῃ, τὰ πολλὰ ἀποθνῄσκει, ἢν πολὺ τὸ ῥεῦμα ἐπιγένηται καὶ νότιον· τὰ γὰρ φλέβια λεπτὰ ἐόντα οὐ δύναται ὑποδέχεσθαι τὸ | φλέγμα ὑπὸ πάχεος καὶ πλήθεος, ἀλλ' ἀποψύχεται καὶ πήγνυται τὸ αἷμα, καὶ οὕτως ἀποθνῄσκει. ἢν δὲ ὀλίγον ᾖ καὶ ἐς ἀμφοτέρας τὰς φλέβας τὸν κατάρροον ποιήσηται, ἢ ἐς τὰς ἐπὶ θάτερα, περιγίνεται ἐπίσημα ἐόντα· ἢ γὰρ στόμα παρέσπασται ἢ ὀφθαλμὸς ἢ χεὶρ ἢ αὐχήν, ὁκόθεν ἂν τὸ φλέβιον πληρωθὲν τοῦ φλέγματος κρατηθῇ καὶ ἀπισχνανθῇ. τούτῳ οὖν τῷ φλεβίῳ ἀνάγκη ἀσθενέστερον εἶναι καὶ ἐνδεέστερον τοῦτο τοῦ σώματος τὸ βλαβέν· ἐς δὲ τὸν πλείω χρόνον ὠφελέει ὡς ἐπὶ τὸ πολύ· οὐ γὰρ ἔτι ἐπίληπτον γίνεται, ἢν ἅπαξ ἐπισημανθῇ, διὰ τόδε· ὑπὸ τῆς ἀνάγκης ταύτης αἱ φλέβες αἱ λοιπαὶ κακοῦνται καὶ μέρος τι συνισχναίνονται, ὥστε τὸν μὲν ἠέρα δέχεσθαι, τοῦ δὲ

376

up and down through the blood, it produces convulsions and pain: hence the kicking. All these effects result when the phlegm, being cold, flows into the blood, which is hot, and makes it freeze and stand still. If the flux is copious and thick, it is immediately fatal, since it overpowers the blood with its coldness and makes it congeal; if the phlegm is weaker, at first it gains the upper hand and impairs respiration, but then after a time, as it spreads through the vessels and mixes with the copious, hot blood, if it is overcome in this way, the vessels take in air, and patients return to their senses.

11. Small children who are seized by this disease usually succumb if the flux is great and the south winds are in the ascendency, for because their small vessels are narrow and cannot admit the phlegm due to its thickness and abundance, their blood is cooled and congeals, bringing death. But if the flux descending is small in amount and goes into vessels either on one side or on both sides, the children survive, but have sequelae: either their mouth is distorted, or an eye, or an arm, or the neck, depending upon where a small vessel on being filled with phlegm was overwhelmed and dried up. This small vessel inevitably causes the injured part of the body to be weaker and significantly impaired. As time passes, however, this injury proves to be beneficial, since there is no further attack in such a child when it has once been marked in this way, because the same cause that injured the first vessel also damages and partly withers other vessels too, so that while they are still capable of receiving air, defluxions of phlegm

φλέγματος τὸν κατάρροον μηκέτι ὁμοίως ἐπικαταρρέειν—ἀσθενέστερα μέντοι ὁμοίως τὰ μέλεα εἰκὸς εἶναι, τῶν φλεβῶν κακωθεισέων. οἷσι δ' ἂν βόρειόν τε καὶ πάνυ ὀλίγον παραρρυῇ καὶ ἐς τὰ δεξιά, ἀσήμως περιγίνονται· κίνδυνος δὲ ξυντραφῆναι καὶ συναυξηθῆναι, ἢν μὴ θεραπευθῶσι τοῖσιν ἐπιτηδείοισιν. τοῖσι μέν νυν παιδίοισιν οὕτω γίνεται, ἢ ὅτι τούτων ἐγγύτατα.

12. (9 L.) Τοὺς δὲ πρεσβυτέρους οὐκ ἀποκτείνει, ὅταν ἐπιγένηται, οὐδὲ διαστρέφει· αἵ τε γὰρ φλέβες εἰσὶ κοῖλαι καὶ αἵματος μεσταὶ θερμοῦ, διότι οὐ δύναται ἐπικρατῆσαι τὸ φλέγμα, οὐδ' ἀποψῦξαι τὸ αἷμα, ὥστε καὶ πῆξαι, ἀλλ' αὐτὸ κρατέεται καὶ καταμείγνυται τῷ αἵματι ταχέως· καὶ οὕτω παραδέχονται αἱ φλέβες τὸν ἠέρα, καὶ τὸ φρόνημα ἐγγίνεται, τά τε σημεῖα
378 τὰ προειρημένα ἧσσον ἐπιλαμβάνει | διὰ τὴν ἰσχύν. τοῖσι δὲ πρεσβυτάτοις ὅταν ἐπιγένηται τοῦτο τὸ νόσημα, διὰ τόδε ἀποκτείνει ἢ παράπληκτον ποιέει, ὅτι αἱ φλέβες κεκένωνται καὶ τὸ αἷμα ὀλίγον τέ ἐστι καὶ λεπτὸν καὶ ὑδαρές. ἢν μὲν οὖν πολὺ καταρρυῇ καὶ χειμῶνος, ἀποκτείνει· ἀπέφραξε[38] γὰρ τὰς ἀναπνοὰς καὶ ἀπέπηξε τὸ αἷμα, ἢν ἐπ' ἀμφότερα ὁ κατάρροος γένηται· ἢν δ' ἐπὶ θάτερα μοῦνον, παράπληκτον ποιέει· οὐ γὰρ δύναται τὸ αἷμα ἐπικρατῆσαι τοῦ φλέγματος λεπτὸν ἐὸν καὶ ψυχρὸν καὶ ὀλίγον, ἀλλ' αὐτὸ κρατηθὲν ἐπάγη, ὥστε ἀκρατέα εἶναι ἐκεῖνα καθ' ἃ τὸ αἷμα διεφθάρη.

[38] ἀπέφραξε Θ: ἀπέπνιξε M

(sc. through them) can no longer proceed to the same extent—these parts are, however, still likely to be weak in the same way because of the damage to their vessels. In cases where the north wind is blowing and the flux is minimal and on the right side, children will survive with no mark, but there is a danger that the disorder will be nourished and augmented as they grow, unless appropriate medical attention is given. This, or something like it, is the situation with children.

12. The sacred disease, when it occurs in older people, is not fatal, nor does it produce distortions in them. This is because their vessels are hollow and filled with warm blood, so that phlegm cannot gain the upper hand or cool the blood making it congeal: rather, the phlegm itself is overcome and quickly mixes with the blood, with the result that the vessels take in air again, and intelligence returns. In these patients the signs I have described are less apparent, due to their strength. When the disease occurs in very old people, it causes death or paralysis because their vessels are empty, and their blood is scanty, thin and watery. Now when a defluxion is violent and occurs in winter, if it involves both sides of the body it will cause death by blocking the respiration and congealing the blood: but if it involves only one side, there will be a paralysis, since the scanty, thin, cold blood will be unable to subdue the phlegm, but be itself overcome and congeal, so that the parts in which the blood is spoiled will lose their strength.

13. (10 L.) Ἐς δὲ τὰ δεξιὰ μᾶλλον καταρρεῖ ἢ ἐς τὰ ἀριστερά, ὅτι αἱ φλέβες ἐπικοιλότεραί εἰσι καὶ πλέονες ἢ ἐν τοῖς ἀριστεροῖς—ἀπὸ γὰρ τοῦ ἥπατος τείνουσι καὶ ἀπὸ τοῦ σπληνός.[39] ἐπικαταρρέει δὲ καὶ ἀποτήκεται τοῖσι μὲν παιδίοισι μάλιστα, οἷς ἂν διαθερμανθῇ ἡ κεφαλὴ ἤν τε ὑπὸ ἡλίου, ἤν τε ὑπὸ πυρός [ἤν τε][40] καὶ ἐξαπίνης φρίξῃ ὁ ἐγκέφαλος· καὶ τότε ἀποκρίνεται τὸ φλέγμα. ἀποτήκεται μὲν γὰρ ἀπὸ τῆς θερμασίης καὶ διαχύσιος τοῦ ἐγκεφάλου· ἄποκρίνεται[41] δὲ ἀπὸ τῆς ψύξιός τε καὶ συστάσιος, καὶ οὕτως ἐπικαταρρέει. τοῖσι μὲν αὕτη ἡ πρόφασις γίνεται, τοῖσι δὲ καὶ ἐπειδὰν ἐξαπίνης μετὰ βόρεια πνεύματα νότος μεταλάβῃ,[42] ξυνεστηκότα τὸν ἐγκέφαλον καὶ ἀσθενέα ὄντα ἔλυσε καὶ ἐχάλασεν, ὥστε πλημυρεῖν τὸ φλέγμα, | καὶ οὕτω τὸν κατάρροον ποιέεται. ἐπικαταρρέει δὲ καὶ ἐξ ἀδήλου φόβου γινομένου, καὶ ἢν δείσῃ βοήσαντός τινος, ἢ μεταξὺ κλαίων μὴ οἷός τε ᾖ τὸ πνεῦμα ταχέως ἀναλαβεῖν, οἷα γίνεται παιδίοισι πολλάκις· ὅ τι δ᾽ ἂν τούτων αὐτῷ γένηται, εὐθὺς ἔφριξε τὸ σῶμα, καὶ ἄφωνος γενόμενος τὸ πνεῦμα οὐχ εἵλκυσεν, ἀλλὰ τὸ πνεῦμα ἠρέμησε, καὶ ὁ ἐγκέφαλος συνέστη, καὶ τὸ αἷμα ἐστάθη, καὶ οὕτως ἀπεκρίθη καὶ ἐπικατερρύη τὸ φλέγμα. τοῖσι μὲν παιδίοισιν αὗται αἱ προφάσιες τῆς ἐπιλήψιός εἰσι τὴν ἀρχήν. τοῖσι δὲ πρεσβύτῃσιν ὁ χειμὼν πολεμιώτατός ἐστιν· ὅταν γὰρ παρὰ πυρὶ πολλῷ διαθερμανθῇ τὴν κεφαλὴν καὶ τὸν ἐγκέφαλον, ἔπειτα ἐν ψύχει γένηται καὶ ῥιγώσῃ, ἢ καὶ ἐκ ψύχεος εἰς ἀλέην ἔλθῃ καὶ παρὰ

380

13. Defluxions occur more often on the right side than on the left, because the vessels there are hollower and more numerous than on the left—that is, the two vessels leading respectively from the liver and the spleen. Fluxes and melting are most frequent in children whose head is overheated by the sun or who sit next to a fire, and whose brain has a sudden chill, which then leads to a secretion of phlegm. At first melting occurs due to the heating and dissolution of the brain, and then secretion due to chilling and condensation; this is how the flux comes about. In some cases, this is the cause, but in others, when subsequent to north winds there is a sudden alteration to south winds, the brain being contracted and weak suddenly relaxes and collapses, so that phlegm floods out and causes a defluxion. Fluxes also arise from fear of the unknown or on being frightened by loud shouting, or if when crying someone cannot quickly catch his breath, as often happens in children. Whenever one of these things happens to someone, his body immediately has a chill, he becomes speechless and cannot draw his breath, which is blocked, his brain contracts, and his blood stands still: this is how phlegm is secreted and flows down. In children these are the original causes of the attacks. In older people, winter is the worst enemy, for when from being beside a large fire a person becomes overheated in his head and brain, and he then moves into the cold and has a chill—or also if he

---

[39] Jones and several other editors delete ἀπὸ γὰρ . . . σπλη-νός as redundant. See Jouanna *Morb. sacr.*, p. 19, n. 4.

[40] Del. Littré [41] ἀποκρ. M: ἐκκρ. Θ

[42] μεταλάβῃ Θ: -βάλλει M

πῦρ πολύ, τὸ αὐτὸ τοῦτο πάσχει, καὶ οὕτως ἐπίληπτος γίνεται κατὰ τὰ προειρημένα. κίνδυνος δὲ πολὺς καὶ ἦρος παθέειν τωὐτὸ τοῦτο, ἢν ἡλιωθῇ ἡ κεφαλή· τὸ δὲ θέρος ἥκιστα, οὐ γὰρ γίνονται μεταβολαὶ ἐξαπιναῖοι. ὅταν δὲ εἴκοσιν ἔτεα παρέλθῃ, οὐκέτι ἡ νοῦσος αὕτη ἐπιλαμβάνει, ἢν μὴ ἐκ παιδίου ξύντροφος ᾖ, ἀλλ' ἢ ὀλίγους ἢ οὐδένα· αἱ γὰρ φλέβες αἵματος μεσταὶ πολλοῦ εἰσίν, καὶ ὁ ἐγκέφαλος συνέστηκε καί ἐστι στιφρός, ὥστ' οὐκ ἐπικαταρρέει ἐς τὰς φλέβας· ἢν δ' ἐπικαταρρυῇ, τοῦ αἵματος οὐ κρατέει, πολλοῦ ἐόντος καὶ θερμοῦ.

14. (11 L.) Ὧι δὲ ἀπὸ παιδίου συνηύξηται καὶ συντέθραπται, ἔθος πεποίηται ἐν τῇσι μεταβολῇσι τῶν 382 πνευμάτων τοῦτο πάσχειν, καὶ ἐπίληπτον | γίνεσθαι ὡς τὰ πολλά, καὶ μάλιστα ἐν τοῖσι νοτίοισιν· ἥ τε ἀπάλλαξις χαλεπὴ γίνεται· ὁ γὰρ ἐγκέφαλος ὑγρότερος γέγονε τῆς φύσιος καὶ πλημμυρεῖ ὑπὸ τοῦ φλέγματος, ὥστε τοὺς μὲν καταρρόους πυκνοτέρους γίνεσθαι, ἐκκριθῆναι δὲ μηκέτι οἷόν τε[43] εἶναι τὸ φλέγμα, μηδὲ ἀναξηρανθῆναι τὸν ἐγκέφαλον, ἀλλὰ διαβεβρέχθαι καὶ εἶναι ὑγρόν. γνοίη δ' ἄν τις τῷδε μάλιστα τοῖσι προβάτοισι τοῖσι καταλήπτοισι γινομένοις ὑπὸ τῆς νούσου ταύτης καὶ μάλιστα τῇσιν αἰξίν· αὗται γὰρ πυκνότατα λαμβάνονται· ἢν διακόψῃς τὴν κεφαλήν, εὑρήσεις τὸν ἐγκέφαλον ὑγρὸν ἐόντα καὶ ὕδρωπος περίπλεον καὶ κακὸν ὄζοντα· καὶ ἐν τούτῳ δηλονότι γνώσῃ ὅτι οὐχ ὁ θεὸς τὸ σῶμα λυμαίνεται, ἀλλ' ἡ νοῦσος. οὕτω δ' ἔχει καὶ τῷ ἀνθρώπῳ· ὁκόταν γὰρ

moves out of the cold into the heat beside a large fire—he will suffer the same effects and have an attack as has been described above. There is also a serious danger of suffering this same thing in the spring, if someone's head is exposed to the sun, but least in the summer since then there are no sudden changes. Once a person has reached twenty years, this disease will no longer attack him, unless it has been habitual since childhood—that is, only in rare cases, or never. This is because by then his vessels are filled with abundant blood, and his brain is contracted and firm, so that there will be no flux into the vessels. And even if there is a flux, the blood is safe from being overwhelmed because of its large amount and warmth.

14. In most cases a person, in whom the disease has grown and been nourished since childhood, becomes used to suffering an attack generally at the changes of the seasons, and especially when there are south winds. Recovery is difficult since his brain has become moister by nature, being flooded with phlegm, making defluxions more frequent: the phlegm can no longer be excreted, nor can the brain be dried, being on the contrary soaked through and moist. This can be understood best from the example of cattle that are affected by attacks from this disease, and especially by the case of goats, who are the most often attacked: if you cut through such an animal's head you will discover the brain to be moist, full of dropsy fluid, and malodorous. This will make you clearly understand that the body is being corrupted not by a god, but by a disease. It is the same in a human being, for once the disease has

43 οἷόν τε M: οἴονται Θ

χρόνος ἐγγένηται τῇ νούσῳ, οὐκέτι ἰήσιμος γίνεται· διεσθίεται γὰρ ὁ ἐγκέφαλος ὑπὸ τοῦ φλέγματος καὶ τήκεται, τὸ δ' ἀποτηκόμενον ὕδωρ γίνεται, καὶ περιέχει[44] τὸν ἐγκέφαλον ἐκτὸς καὶ περικλύζει· καὶ διὰ τοῦτο πυκνότερον ἐπίληπτοι γίνονται καὶ ῥᾷον. διὸ δὴ πολυχρόνιος ἡ νοῦσος, <καὶ>[45] ὅτι τὸ περιρρέον λεπτόν ἐστιν ὑπὸ πολυπληθείης, καὶ εὐθὺς κρατέεται ὑπὸ τοῦ αἵματος καὶ διαθερμαίνεται.

15. (12 L.) Ὅσοι δὲ ἤδη ἐθάδες εἰσὶ τῇ νούσῳ, προγινώσκουσιν ὅταν μέλλωσι ληφθήσεσθαι, καὶ φεύγουσιν ἐκ τῶν ἀνθρώπων, ἢν μὲν ἐγγὺς ᾖ αὐτῷ τὰ οἰκία, οἴκαδε, εἰ δὲ μή, ἐς τὸ ἐρημότατον, ὅπῃ μέλλουσιν αὐτὸν ἐλάχιστοι ὄψεσθαι πεσόντα, εὐθύς τε ἐγκαλύπτεται· τοῦτο δὲ ποιέει ὑπ' αἰσχύνης τοῦ πάθεος καὶ οὐχ ὑπὸ φόβου, ὡς οἱ πολλοὶ νομίζουσι, τοῦ δαιμονίου. τὰ δὲ παιδάρια τὸ μὲν πρῶτον πίπτουσιν ὅπῃ ἂν τύχωσιν ὑπὸ ἀηθίης· ὅταν δὲ πολλάκις κατάληπτοι | γένωνται, ἐπειδὰν προαίσθωνται, φεύγουσι παρὰ τὰς μητέρας ἢ παρὰ ἄλλον ὅντινα μάλιστα γινώσκουσιν, ὑπὸ δέους καὶ φόβου τῆς πάθης· τὸ γὰρ αἰσχύνεσθαι οὔπω γινώσκουσιν. 384

16. (13 L.) Ἐν δὲ τῇσι μεταβολῇσι τῶν πνευμάτων διὰ τάδε φημὶ ἐπιλήπτους γίνεσθαι, καὶ μάλιστα τοῖσι νοτίοισιν, ἔπειτα τοῖσι βορείοισιν, ἔπειτα τοῖσι λοιποῖσι πνεύμασι· ταῦτα γὰρ τῶν λοιπῶν πνευμάτων ἰσχυρότατά ἐστι καὶ ἀλλήλοις ἐναντιώτατα κατὰ τὴν στάσιν καὶ κατὰ τὴν δύναμιν. ὁ μὲν γὰρ βορέης ξυνίστησι τὸν ἠέρα καὶ τὸ θολερόν τε καὶ τὸ νοτῶδες[46]

become chronic it is no longer curable, since the brain has been corroded by the phlegm and melted, and what melts away turns to liquid, which surrounds the brain on the outside and inundates it completely: this is why people suffer attacks more frequently and with less apparent cause. As a result of this the disease lasts a long time, and also because the flux being added is thin, on account of its great amount, and therefore quickly dominated by the blood and warmed up.

15. Patients already used to the disease know in advance when they are about to have an attack, and leave the presence of others. If their home is close, they go home, but if not, they flee to deserted places where the fewest people will see them falling, and they immediately cover themselves up. This they do out of shame for their disease, and not, as many people think, from fear of something divine. But young children at first fall wherever they happen to be, from their lack of experience, but then after they have had numerous attacks, when they sense one in advance, they flee to their mothers or whomever else they know best, from fear and terror of the disease, since they do not yet know what shame is.

16. This is why, in my opinion, people have attacks when the winds change, in particular to the south wind, next to the north wind, and then the other winds. These two (sc. south and the north) winds are the strongest of all, and most opposed in their direction and character: the north wind compacts the air and expels from it whatever

---

[44] περιέχει M: περιχέει Θ [45] Add. Jouanna
[46] νοτῶδες Θ: νεφῶδες M

ἐκκρίνει καὶ λαμπρόν τε καὶ διαφανέα ποιέει. κατὰ δὲ τὸν αὐτὸν τρόπον καὶ τἆλλα πάντα ἐκ τῆς θαλάσσης ἀρξάμενα καὶ τῶν ἄλλων ὑδάτων· ἐκκρίνει γὰρ ἐξ ἁπάντων τὴν νοτίδα καὶ τὸ δνοφερόν, καὶ γὰρ ἐξ αὐτῶν τῶν ἀνθρώπων, διὸ καὶ ὑγιηρότατός ἐστι τῶν ἀνέμων. ὁ δὲ νότος τἀντία τούτῳ ἐργάζεται· πρῶτον μεν ἄρχεται τὸν ἠέρα συνεστηκότα κατατήκειν καὶ διαχέειν, καθότι καὶ οὐκ εὐθὺς πνεῖ μέγας, ἀλλὰ γαληνίζει[47] πρῶτον, ὅτι οὐ δύναται ἐπικρατῆσαι τοῦ ἠέρος αὐτίκα, τοῦ πρόσθεν πυκνοῦ τε ἐόντος καὶ συνεστηκότος, ἀλλὰ τῷ χρόνῳ διαλύει. τὸ δ᾽ αὐτὸ τοῦτο καὶ τὴν γῆν ἐργάζεται καὶ τὴν θάλασσαν καὶ ποταμοὺς καὶ κρήνας καὶ φρέατα καὶ ὅσα φύεται καὶ ἐν οἷς τι ὑγρόν ἔνεστιν· ἔστι δ᾽ ἐν παντί, ἐν τῷ μὲν πλέον, ἐν τῷ δ᾽ ἔλασσον. ἅπαντα δὲ ταῦτα αἰσθάνεται τοῦ πνεύματος τούτου, καὶ ἔκ τε λαμπρῶν δνοφώδεα γίνεται, καὶ ἐκ ψυχρῶν θερμά, καὶ ἐκ ξηρῶν νοτώδεα.
386 ὅσα δ᾽ ἐν | οἰκήμασι κεράμεα ἢ κατὰ γῆς ἐστι μεστὰ οἴνου ἢ ἄλλου τινὸς ὑγροῦ, πάντα ταῦτα αἰσθάνεται τοῦ νότου καὶ διαλλάσσει τὴν μορφὴν ἐς ἕτερον εἶδος. τόν τε ἥλιον καὶ τὴν σελήνην καὶ τἆλλα ἄστρα πολὺ ἀμβλυωπότερα καθίστησι τῆς φύσιος. ὅτε οὖν καὶ τούτων οὕτω μεγάλων ἐόντων καὶ ἰσχυρῶν τοσοῦτον ἐπικρατέει, καὶ τὸ σῶμα ποιέει αἰσθάνεσθαι καὶ μεταβάλλειν. ἐν[48] τῶν ἀνέμων τούτων τῇσι μεταλλαγῇσιν, ἀνάγκη τοῖσι μὲν νοτίοισι λύεσθαί τε καὶ φλυδᾶν τὸν ἐγκέφαλον καὶ τὰς φλέβας χαλαρωτέρας γίνεσθαι, τοῖσι δὲ βορείοισι ξυνίστασθαι τὸ ὑγιη-

is turbid and moist, making it bright and clear. In fact it has this same effect on everything, beginning from the sea and other bodies of water: it expels the damp and dull component from everything, even from human beings themselves, and for this reason it is the healthiest of winds. The south wind has an effect opposite to this: first, it begins by melting and dispersing air that is condensed, and because its force is not immediately strong it first brings fair weather, being unable to gain control at once over the air that was previously dense and compressed; with time, however, it does dissolve the air. It also has the same effect on the earth, the sea, rivers, springs, and wells, as well as on everything that is alive and contains moisture, and there is moisture in all things, more in some, less in others. All such things are sensitive to the south wind, and change from being bright to being dull, from cold to warm, from dry to moist. Pottery vessels kept in houses or underground which contain wine or any other liquid all sense the south wind, and change their shape to another form. It also makes the sun, the moon, and the other stars much duller than in their natural state. Now since this wind has so much influence over such large and powerful things, it will certainly also make the (sc. human) body perceive it and change. In the changes of these winds, under the influence of the southerlies, the brain must be relaxed and moistened, and the vessels must become slack. From the northerlies, by contrast, the healthiest part of the brain

---

[47] γαληνίζει Θ: λαγανίζει M
[48] ἐν Θ: ἐκ M

ρότατον τοῦ ἐγκεφάλου, τὸ δὲ νοσηλότατον καὶ ὑγρότατον ἐκκρίνεσθαι καὶ περικλύζειν ἔξωθεν, καὶ οὕτω τοὺς καταρρόους ἐπιγίνεσθαι ἐν τῇσι μεταβολῇσι τούτων τῶν πνευμάτων. οὕτως αὕτη ἡ νοῦσος γίνεται καὶ θάλλει ἀπὸ τῶν προσιόντων τε καὶ ἀπιόντων, καὶ οὐδέν ἐστιν ἀπορωτέρη τῶν ἄλλων οὔτε ἰῆσθαι οὔτε γνῶναι, οὐδὲ θειοτέρη ἢ αἱ ἄλλαι.

17. (14 L.) Εἰδέναι δὲ χρὴ τοὺς ἀνθρώπους, ὅτι ἐξ οὐδενὸς ἡμῖν αἱ ἡδοναὶ γίνονται καὶ εὐφροσύναι καὶ γέλωτες καὶ παιδιαὶ ἢ ἐντεῦθεν, ὅθεν καὶ λῦπαι καὶ ἀνίαι καὶ δυσφροσύναι καὶ κλαυθμοί. καὶ τούτῳ φρονέομεν μάλιστα καὶ νοέομεν καὶ βλέπομεν καὶ ἀκούομεν καὶ διαγινώσκομεν τά τε αἰσχρὰ καὶ τὰ καλὰ καὶ τὰ κακὰ καὶ τἀγαθὰ καὶ ἡδέα καὶ ἀηδέα, τὰ μὲν νόμῳ διακρίνοντες, τὰ δὲ τῷ ξυμφέροντι αἰσθανόμενοι. [τῷ δὲ καὶ τὰς ἡδονὰς καὶ τὰς ἀηδίας τοῖς καιροῖς διαγινώσκοντες, καὶ οὐ ταὐτὰ ἀρέσκει ἡμῖν.][49] τῷ δ' αὐτῷ τούτῳ καὶ μαινόμεθα καὶ παραφρονέομεν, καὶ δείματα καὶ φόβοι παρίστανται ἡμῖν | τὰ μὲν νύκτωρ, τὰ δὲ καὶ μεθ' ἡμέρην, καὶ ἐνύπνια[50] καὶ πλάνοι ἄκαιροι, καὶ φροντίδες οὐχ ἱκνεύμεναι, καὶ ἀγνωσίαι τῶν καθεστεώτων καὶ ἀηθίαι.[51] καὶ ταῦτα πάσχομεν ἀπὸ τοῦ ἐγκεφάλου πάντα, ὅταν οὗτος μὴ ὑγιαίνῃ, ἀλλ' ἢ θερμότερος τῆς φύσιος γένηται ἢ ψυχρότερος ἢ ὑγρότερος ἢ ξηρότερος, ἤ τι ἄλλο πεπόνθῃ πάθος παρὰ τὴν

388

[49] Del. Jones as being a gloss; cf. Jouanna *Morb. sacr.*, p. 26, n. 1

must contract, and the moistest and most diseased part must be excreted and washed away from the outside, so that defluxions result in the changes of these winds. In this way the sacred disease is engendered and advanced by what passes into and out of the body, and it is no more irremediable than other diseases, or impossible to understand, nor is it diviner than the rest.

17. People must know that from the brain, and from the brain alone, come our pleasures, our joys, our laughter, and our jests, as well as our sorrows, our troubles, our griefs, and our tears. With it in particular we think, we contemplate, we look, we listen, we judge between which things are ugly and which beautiful, which are bad and which good, which are pleasant and which unpleasant, deciding in some cases on the basis of convention, judging in other cases according to what is advantageous [and sometimes discerning what is pleasure and what displeasure according to what seems opportune, since it is not always the same thing that appeals to us]. By this same (sc. brain) we are mad, we become deranged, we are visited by fears and terrors—some by night and others by day—we have nightmares, we make inopportune mistakes, we are the prey of imaginary worries, we become unable to grasp reality, and we make strange. All these things we suffer as a result of the brain becoming unhealthy, due to the effects of excessive heat, cold, moistness or dryness, or suffering some other change against nature, to which it

---

[50] ἐνύπνια M: ἀγρυπνίαι Θ

[51] ἀηθίαι Θ: ἀηθίη καὶ ἀπειρίη M: λήθη καὶ ἀπορίη Co

φύσιν ὃ μὴ ἐώθει. καὶ μαινόμεθα μὲν ὑπὸ ὑγρότητος· ὅταν γὰρ ὑγρότερος τῆς φύσιος ᾖ, ἀνάγκη κινέεσθαι, κινευμένου δὲ μήτε τὴν ὄψιν ἀτρεμίζειν μήτε τὴν ἀκοήν, ἀλλ' ἄλλοτε ἄλλα ὁρᾶν καὶ ἀκούειν, τήν τε γλῶσσαν τοιαῦτα διαλέγεσθαι οἷα ἂν βλέπῃ τε καὶ ἀκούῃ ἑκάστοτε· ὅσον δ' ἂν ἀτρεμίσῃ ὁ ἐγκέφαλος χρόνον, τοσοῦτον καὶ φρονέει ὁ ἄνθρωπος.

18. (15 L.) Γίνεται δὲ ἡ διαφθορὴ τοῦ ἐγκεφάλου ὑπὸ φλέγματος καὶ χολῆς· γνώσει δὲ ἑκάτερα ὧδε· οἱ μὲν ὑπὸ φλέγματος μαινόμενοι ἥσυχοί τέ εἰσι καὶ οὐ βοηταὶ οὐδὲ θορυβώδεες, οἱ δὲ ὑπὸ χολῆς κεκρᾶκταί τε καὶ κακοῦργοι καὶ οὐκ ἀτρεμαῖοι, ἀλλ' αἰεί τι ἄκαιρον δρῶντες. ἢν μὲν οὖν ξυνεχέως μαίνωνται, αὗται αἱ προφάσιές εἰσιν· ἢν δὲ δείματα καὶ φόβοι παριστῶνται, ὑπὸ μεταστάσιος τοῦ ἐγκεφάλου· μεθίσταται δὲ θερμαινόμενος· θερμαίνεται δὲ ὑπὸ τῆς χολῆς, ὅταν ὁρμήσῃ ἐπὶ τὸν ἐγκέφαλον κατὰ τὰς φλέβας τὰς αἱματίτιδας ἐκ τοῦ σώματος· καὶ ὁ φόβος παρέστηκε μέχρι ἀπέλθῃ πάλιν ἐς τὰς φλέβας καὶ τὸ σῶμα· ἔπειτα πέπαυται. ἀνιᾶται δὲ καὶ ἀσᾶται παρὰ καιρὸν ψυχομένου τοῦ ἐγκεφάλου καὶ ξυνισταμένου παρὰ τὸ ἔθος· τοῦτο δὲ ὑπὸ φλέγματος πάσχει· ὑπ' αὐτοῦ δὲ τοῦ πάθεος καὶ ἐπιλήθεται. ἐκ νυκτῶν δὲ βοᾷ καὶ κέκραγεν, ὅταν ἐξαπίνης | ὁ ἐγκέφαλος διαθερμαίνηται· τοῦτο δὲ πάσχουσιν οἱ χολώδεες, οἱ δὲ φλεγματώδεες οὔ· διαθερμαίνεται δὲ καὶ ἐπὴν τὸ αἷμα ἐπέλθῃ ἐπὶ τὸν ἐγκέφαλον πολὺ καὶ ἐπιζέσῃ. ἔρχεται δὲ κατὰ τὰς φλέβας πολὺ τὰς προειρημένας, ὅταν τυγχάνῃ

390

is not accustomed. We become mad from moistness, since when the brain becomes moister than is natural, it is compelled to move, and in moving it disturbs seeing and hearing, so that we see and hear one thing at one moment, and another at another moment, and the tongue then reports what we are seeing and hearing at these different times. But as long as the brain remains still, a person can think.

18. Impairment of the brain is caused by phlegm and bile, and each of the two you will recognize by the following. Patients whose madness is caused by phlegm are peaceful, keep quiet, and stay still, while those who are mad because of bile are obstreperous, troublesome, and restless, always doing improper things. These are the causes of permanent madness, but if temporary fears and terrors assail people, their brain is being altered by being heated when bile rushes to it from the body through the blood vessels; and the fear lasts until the bile returns to the vessels and the body, then stopping. A patient suffers untimely distress and loathing when his brain is cooled and contracted due to phlegm in an unaccustomed way, and in this condition there is also a loss of memory. A person shouts and cries out in the night when his brain is suddenly heated, an affection associated with bile, but not phlegm. The brain can also be heated when a large amount of blood moves into it and effervesces; this blood passes through the vessels mentioned above, if a person happens

ὥνθρωπος ἐνύπνιον ὁρῶν φοβερὸν καὶ ἐν πόνῳ[52] ᾖ· ὥσπερ οὖν καὶ ἐγρηγορότι τότε μάλιστα τὸ πρόσωπον φλογιᾷ, καὶ οἱ ὀφθαλμοὶ ἐρεύθονται, ὅταν φοβῆται, καὶ ἡ γνώμη ἐπινοέῃ τι κακὸν ἐργάσασθαι, οὕτω καὶ ἐν τῷ ὕπνῳ πάσχει. ὅταν δ' ἐπέγρηται καὶ καταφρονήσῃ καὶ τὸ αἷμα πάλιν σκεδασθῇ ἐς τὰς φλέβας πέπαυται.

19. (16 L.) Κατὰ ταῦτα νομίζω τὸν ἐγκέφαλον δυναμιν ἔχειν πλείστην ἐν τῷ ἀνθρώπῳ· οὗτος γὰρ ἡμῖν ἐστι τῶν ἀπὸ τοῦ ἠέρος γινομένων ἑρμηνεύς, ἢν ὑγιαίνων τυγχάνῃ· τὴν δὲ φρόνησιν αὐτῷ[53] ὁ ἀὴρ παρέχεται. οἱ δ' ὀφθαλμοὶ καὶ τὰ ὦτα καὶ ἡ γλῶσσα καὶ αἱ χεῖρες καὶ οἱ πόδες οἷα ἂν ὁ ἐγκέφαλος γινώσκῃ, τοιαῦτα ὑπηρετοῦσι·[54] γίνεται γὰρ ἐν ἅπαντι τῷ σώματι τῆς φρονήσιος, τέως ἂν μετέχῃ τοῦ ἠέρος. ἐς δὲ τὴν ξύνεσιν ὁ ἐγκέφαλός ἐστιν ὁ διαγγέλλων· ὅταν γὰρ σπάσῃ τὸ πνεῦμα ὥνθρωπος ἐς ἑωυτόν, ἐς τὸν ἐγκέφαλον πρῶτον ἀφικνεῖται, καὶ οὕτως ἐς τὸ λοιπὸν σῶμα σκίδναται ὁ ἀήρ, καταλελοιπὼς ἐν τῷ ἐγκεφάλῳ ἑωυτοῦ τὴν ἀκμὴν καὶ ὅ τι ἂν ᾖ φρόνιμόν τε καὶ γνώμην ἔχον· εἰ γὰρ ἐς τὸ σῶμα πρῶτον ἀφικνεῖτο καὶ ὕστερον ἐς τὸν ἐγκέφαλον, ἐν τῇσι σαρξὶ
392 καὶ ἐν τῇσι φλεψὶ καταλελοιπὼς τὴν | διάγνωσιν ἐς τὸν ἐγκέφαλον ἂν ᾔει θερμὸς ἐὼν καὶ οὐκ ἀκραιφνής, ἀλλὰ ἐπιμεμιγμένος τῇ ἰκμάδι τῇ ἀπό τε τῶν σαρκῶν καὶ τοῦ αἵματος, ὥστε μηκέτι εἶναι ἀκριβής.

20. (17 L.) Διότι φημὶ τὸν ἐγκέφαλον εἶναι τὸν ἑρμηνεύοντα τὴν ξύνεσιν. αἱ δὲ φρένες ἄλλως ὄνομα

to have a terrifying nightmare and suffer pain. Thus, just as a waking person's face at that time in particular becomes flushed and his eyes turn red, when he is suffering an attack of fear and is contemplating doing some untoward thing, the same happens in a sleeping person. Then, when he awakes and comes to his senses, this stops as the blood disperses back into the vessels.

19. For these reasons, I believe that the brain possesses the greatest power in a person: if it is healthy, it is our interpreter of the things coming from the air, since air provides it with intelligence. The eyes, ears, tongue, hands and feet also carry out what the brain thinks, since there is intention in every part of the body as long as it possesses air. The brain is the messenger of intelligence: when a person draws in breath, this arrives first in the brain, and then, after leaving its highest part, which is intelligent and possesses thought, in the brain, the air spreads out into the rest of the body. For if (sc. the air) arrived first in the body, and only later in the brain, after having left its intelligence in the tissues and vessels, it would arrive in the brain hot and impure, mixed with moisture from the flesh and blood, so that it would no longer possess the property of precise discernment.

20. This is why I call the brain the interpreter of understanding. The diaphragm (*phrenes*) received its name

---

52 *πόνῳ* Θ: *τῷ φόβῳ* M

53 *αὐτῷ* M: om. Θ

54 *ὑπηρετοῦσι* M: *πρήσσουσι* Θ

ἔχουσι τῇ τύχῃ κεκτημένον καὶ τῷ νόμῳ, τῷ δ' ἐόντι οὔκ, οὐδὲ[55] τῇ φύσει, οὐδὲ οἶδα ἔγωγε τίνα δύναμιν ἔχουσιν αἱ φρένες ὥστε νοέειν τε καὶ φρονέειν, πλὴν ἤν τι ὥνθρωπος ὑπερχαρῇ ἐξ ἀδοκήτου ἢ ἀνιαθῇ, πηδῶσι καὶ ἅλσιν παρέχουσιν ὑπὸ λεπτότητος καὶ ὅτι ἀνατέτανται μάλιστα ἐν τῷ σώματι, καὶ κοιλίην οὐκ ἔχουσι ἐς ἥντινα χρὴ δέξασθαι ἢ ἀγαθὸν ἢ κακὸν προσπῖπτον, ἀλλ' ὑπὸ ἀμφοτέρων τούτων τεθορύβηνται διὰ τὴν ἀσθενείην τῆς φύσιος· ἐπεὶ αἰσθάνονταί γε οὐδενὸς πρότερον τῶν ἐν τῷ σώματι ἐόντων, ἀλλὰ μάτην τοῦτο τὸ ὄνομα ἔχουσι καὶ τὴν αἰτίην, ὥσπερ τὰ πρὸς τῇ καρδίῃ ὦτα καλέεται, οὐδὲν ἐς τὴν ἀκοὴν ξυμβαλλόμενα. λέγουσι δέ τινες ὡς καὶ φρονέομεν τῇ καρδίῃ καὶ τὸ ἀνιώμενον τοῦτό ἐστι καὶ τὸ φροντίζον· τὸ δὲ οὐχ οὕτως ἔχει, ἀλλὰ σπᾶται μὲν ὥσπερ αἱ φρένες καὶ μᾶλλον διὰ ταύτας τὰς αἰτίας· ἐξ ἅπαντος τοῦ σώματος φλέβες ἐς αὐτὴν τείνουσι, καὶ ξυγκλείσασα ἔχει ὥστε αἰσθάνεσθαι, ἤν τις πόνος ἢ τάσις γένηται τῷ ἀνθρώπῳ· ἀνάγκη δὲ καὶ ἀνιώμενον φρίσσειν τε τὸ σῶμα καὶ συντείνεσθαι, καὶ ὑπερχαίροντα τωὐτὸ τοῦτο πάσχειν· διότι ἡ καρδίη αἰσθάνεταί τε 394 μάλιστα | καὶ αἱ φρένες· τῆς μέντοι φρονήσιος οὐδετέρῳ μέτεστιν, ἀλλὰ πάντων τούτων αἴτιος ὁ ἐγκέφαλός ἐστιν. ὡς οὖν καὶ τῆς φρονήσιος τοῦ ἠέρος πρῶτος αἰσθάνεται τῶν ἐν τῷ σώματι ἐόντων, οὕτω καὶ ἤν τις μεταβολὴ ἰσχυρὴ γένηται ἐν τῷ ἠέρι ὑπὸ τῶν ὡρέων, καὶ αὐτὸς ἑωυτοῦ διάφορος γίνεται [ἐν τῷ ἠέρι][56] ὁ ἐγκέφαλος πρῶτος αἰσθάνεται. διότι καὶ τὰ

for no good reason, but by chance and convention, and not from either reality or nature. Nor am I aware of any potency the diaphragm possesses by which it could apprehend and think (*phronein*), except in the situation where a person suffered an unexpected excess of joy or grief, and his diaphragm pounded and leaped up because of its being the thinnest and most tightly stretched part of the body. The diaphragm has no cavity into which any good or bad thing invading it could be received, but it is susceptible to disturbance by these two (sc. emotions) because of the weakness of its structure. Since it perceives nothing in advance of the other parts in the body, it has its name for no good reason or cause, just as the auricles (ears) next to the heart have their name in spite of the fact that they contribute nothing to hearing. Some people claim that we think with our heart, and that it can be vexed and worried, but this is not so, but rather it is agitated the way the diaphragm is, but even more, for the following reasons. Vessels to the heart lead there from the whole body, and the heart, grasping these closely, has the property of sensing when a person suffers any pain or tension. From this it follows necessarily that when a person is grieved his body shivers and contracts, and the same when he is overjoyed. This is why the heart is the most sensitive part, along with the diaphragm, but neither of them has any share of thought, whose cause is entirely in the brain. Now just as the brain senses the air's thought first of the parts of the body, if some violent change occurs in the air on account of the seasons and it becomes different from how it was, the brain senses this first. Thus, I assert that the diseases

---

[55] οὔκ, οὐδὲ Cornarius in marg.: οὐδὲ M: οὐ Θ
[56] Del. Ermerins

νοσήματα ἐς αὐτὸν ἐμπίπτειν φημὶ ὀξύτατα καὶ μέγιστα καὶ θανατωδέστατα καὶ δυσκριτώτατα τοῖς ἀπείροισιν.

21. (18 L.) *Αὕτη δὲ ἡ νοῦσος ἡ ἱερὴ καλεομένη ἀπο τῶν αὐτῶν προφασίων γίνεται ἀφ᾽ ὧν καὶ αἱ λοιπαὶ ἀπὸ τῶν προσιόντων καὶ ἀπιόντων, καὶ ψύχεος καὶ ἡλίου καὶ πνευμάτων μεταβαλλομένων τε καὶ οὐδέποτε ἀτρεμιζόντων. ταῦτα δ᾽ ἐστὶ θεῖα, ὥστε μὴ δεῖν ἀποκρίνοντα τὸ νόσημα θειότερον τῶν λοιπῶν νομίσαι, ἀλλὰ πάντα θεῖα καὶ πάντα ἀνθρώπινα· φύσιν δὲ ἕκαστον ἔχει καὶ δύναμιν ἐφ᾽ ἑωυτοῦ, καὶ οὐδὲν ἄπορόν ἐστιν οὐδὲ ἀμήχανον· ἀκεστά τε τὰ πλεῖστά ἐστι τοῖς αὐτοῖσι τούτοισιν ἀφ᾽ ὧν καὶ γίνεται. ἕτερον γὰρ ἑτέρῳ τροφή ἐστι, τῷ*[57] *δὲ καὶ κάκωσις. τοῦτο οὖν δεῖ τὸν ἰητρὸν ἐπίστασθαι, ὅπως τὸν καιρὸν διαγινώσκων ἑκάστου τῷ μὲν ἀποδώσει τὴν τροφὴν καὶ αὐ-* 396 *ξήσει, τῷ δὲ ἀφαιρήσει καὶ μειώσει. χρὴ γὰρ καὶ* | *ἐν ταύτῃ τῇ νούσῳ καὶ ἐν τῇσιν ἄλλῃσιν ἁπάσῃσι μὴ αὔξειν τὰ νοσήματα, ἀλλὰ τρύχειν προσφέροντα τῇ νούσῳ τὸ πολεμιώτατον ἑκάστῃ καὶ μὴ τὸ σύνηθες· ὑπὸ μὲν γὰρ τῆς συνηθείης θάλλει καὶ αὔξεται, ὑπὸ δὲ τοῦ πολεμίου φθίνει τε καὶ ἀμαυροῦται. ὅστις δ᾽ ἐπίσταται ἐν ἀνθρώποισι ξηρὸν καὶ ὑγρὸν ποιέειν, καὶ ψυχρόν καὶ θερμόν, ὑπὸ διαίτης, οὗτος καὶ ταύτην τὴν νοῦσον ἰῷτο ἄν, εἰ τοὺς καιροὺς διαγινώσκοι τῶν συμφερόντων, ἄνευ καθαρμῶν καὶ μαγίης.*[58]

[57] *τῷ* Cornarius in marg.: *τὸ* M: *τότε* Θ
[58] Add. *καὶ πάσης τῆς τοιαύτης βαναυσίης* M

which attack the brain are the most acute, serious, and dangerous, and the most difficult for the incompetent to judge.

21. This disease called sacred arises from the same causes as all the rest: from what comes to the body and what leaves it, and from cold, the sun, and from winds constantly changing and never being at rest. These things are divine, so that there is no need to set this disease apart and consider it more divine than any others, but to accept that they are all divine, and all human: each has a nature and a potency of its own, and none is hopeless or beyond management: most are curable by the same things from which they arose. Some agents are nutriment for some diseases, but to other diseases deleterious. The physician must understand this in order to recognize the opportune moment in every case, when to give a nutriment and increase it, when to withdraw a nutriment and decrease it. For both in this disease and all others you must avoid increasing the diseases, but wear them down by administering the thing most deleterious to each disease, rather than what it is accustomed to; for from what is accustomed a disease flourishes and grows, while from what is deleterious it declines and withers away. Anyone who understands how to produce dryness, moistness, coldness and warmth in people by means of diet will be able to cure this disease too, if he is able to recognize the opportune moments for treatment—and this without recourse to purifications and magic.[2]

[2] M adds "and all this kind of nonsense."

THE ART

INTRODUCTION

The Art is a rhetorical discourse both justifying the medical art directly and defending it against arguments of its unnamed critics. Its chapters are as follows:

1	Introductory statement on disparagers and deniers of the arts in general.
2	Argument that all the known arts exist.
3	Narrowing the focus of this speech to the art of medicine.
4/7	Refutation of the argument: some patients die although they employ medicine.
5	Refutation of the argument: some survive although they do not employ medicine.
6	Argument that medicine has many effective methods and practices.
8	Refutation of the argument: physicians refuse to take on incurable cases.
9	Argument that diseases with observable lesions can be understood and treated.
10	Explanation of anatomical parts and processes in diseases with internal lesions.
11	Argument that occult diseases can be understood, although slowly, by the mind.

12 Argument that all the arts, including medicine, require time and carefulness.
13 Argument that medicine has methods of forcing nature to reveal occult diseases.
14 Confirmation of medicine's strengths, limits, and basis in reality.

The author's assumptions about the universe, the body, and disease, and his methods of argumentation, suggest that he belongs to the last decades of the fifth century BC, and to the milieu of the sophists. Whether he is in fact a sophist himself, or a physician, is not agreed upon by scholars.[1]

Erotian's census of Hippocratic texts includes the title περὶ τέχνης, and six entries in his *Glossary* are derived from it.[2] The work seems unknown to Galen, but other ancient texts show knowledge of the text,[3] including the pseudo-Galenic *Medical Definitions* (= *Med. Def.*).[4] Latin translations of parts of *The Art* are present in several manuscripts dating from the ninth to the fourteenth centuries, and Latin translations of two chapters made by Andreas Brentius of Padua are printed in the volume *Hippocratis De natura hominis. . . . Demonstratio quod artes sunt*

[1] See Gomperz, pp. 3–35; Jones, vol. 2, pp. 186–88; Jouanna *Art*, pp. 175–83; Ducatillon, pp. 43–83.

[2] See *Erotianstudien*, p. 341f.: A63, ἀκέσιας; Θ4, θαλάμαι; M14, μωμητέον; N9, νηδύν; Σ20, σοφίην; Υ10, ὕποφρον.

[3] Cf. *Testimonien* vol. I, pp. 185–88; vol. II,1, p. 166; vol. II,2, p. 117f.; vol. III, p. 166f.

[4] Cf. Jouanna *Art*, pp. 208–10.

(= ch. 2); *Invectiva in obtrectatores medicinae* (= ch. 1) (Rome, ca. 1483–1490).[5]

The treatise is contained in the two independent Greek manuscripts M and A, and for chapter 5 Vaticanus Urbinas Graecus 64 (XII c.) (= Urb) is a further independent witness. *The Art* is present in the collected editions and translations of the Corpus, including Zwinger, Mack, Daremberg, Heiberg, Chadwick (under the title *The Science of Medicine*), and Diller *Schr.*, and important special studies devoted to it are:

Gomperz, Th. *Die Apologie der Heilkunst. Eine griechische Sophistenrede des fünften vorchristlichen Jahrhunderts.* Vienna, 1890. (2nd rev. ed., Leipzig, 1910). (= Gomperz)

Diels, H. "Hippokratische Forschungen IV." *Hermes* 48 (1913): 378–407. (= Diels)

Jouanna, J. *Hippocrate. . . . De l' Art*. Budé V (1). Paris, 1988. (= Jouanna *Art*)

The chapter numbering in the present edition follows Gomperz and Jones, with Littré's numbers in brackets where they are different.

[5] Cf. Kibre, pp. 91–93; Schullian, p. 107.

ΠΕΡΙ ΤΕΧΝΗΣ

VI 2 Littré

1. Εἰσί τινες οἳ τέχνην πεποίηνται τὸ τὰς τέχνας αἰσχροεπεῖν, ὡς μὲν οἴονται, οὐ τοῦτο διαπρησσόμενοι ὃ ἐγὼ λέγω, ἀλλ' ἱστορίης οἰκείης ἐπίδειξιν ποιεύμενοι. ἐμοὶ δὲ τὸ μέν τι τῶν μὴ εὑρημένων ἐξευρίσκειν, ὅ τι καὶ εὑρεθὲν κρέσσον ἢ ἀνεξεύρετον, ξυνέσιος δοκέει ἐπιθύμημά τε καὶ ἔργον εἶναι, καὶ τὸ τὰ ἡμίεργα ἐς τέλος ἐξεργάζεσθαι ὡσαύτως· τὸ δὲ λόγων οὐ καλῶν τέχνῃ τὰ τοῖς ἄλλοις εὑρημένα αἰσχύνειν προθυμέεσθαι, ἐπανορθοῦντα μὲν μηδέν, διαβάλλοντα δὲ τὰ τῶν εἰδότων πρὸς τοὺς μὴ εἰδότας ἐξευρήματα, οὐκέτι συνέσιος δοκέει ἐπιθύμημά τε καὶ ἔργον εἶναι, ἀλλὰ κακαγγελίη μᾶλλον φύσιος ἢ ἀτεχνίη· μούνοισι γὰρ δὴ τοῖσιν ἀτέχνοισιν ἡ ἐργασίη αὕτη ἁρμόζει, φιλοτιμεομένων μέν, οὐδαμὰ δὲ δυναμένων κακίῃ ὑπουργέειν ἐς τὸ τὰ τῶν πέλας ἔργα ἢ ὀρθὰ ἐόντα διαβάλλειν, ἢ οὐκ ὀρθὰ μωμέεσθαι. τοὺς μὲν οὖν ἐς τὰς ἄλλας τέχνας τούτῳ τῷ τρόπῳ ἐμπίπτοντας, οἷσι μέλει τε, καὶ ὧν μέλει, οἱ δυνάμενοι κωλυόντων· ὁ δὲ παρεὼν λόγος τοῖσιν ἐς ἰητρικὴν οὕτως ἐπιπορευομένοις[1] ἐναντιώσεται, θρασυνόμενος μὲν διὰ τούτους οὓς

THE ART

1. There are people who have made an art of disparaging the arts, although they deny they are doing what I claim, but believe they are making a display of their own cleverness. Now in my opinion, to discover something among things unknown, which would be better if it were discovered rather than left undiscovered, is the (sc. proper) ambition and task of intelligence, and equally so to bring things that are half-accomplished to completion. But to strive, by the art of dishonest arguments, to discredit the discoveries made by other people, correcting nothing, but deriding to the uninformed what the informed have discovered, can no longer be regarded as the ambition and task of intelligence, but is rather a bad sign of character even more than a lack of art. For this practice would appeal only to those lacking in art, who, being ambitious but completely incapable, indulge their meanness by misrepresenting the works of others that are correct, and finding fault with those that are incorrect. Now those who attack the other arts in this way, let competent people rebut who care about these things and are familiar with them. The present discourse is aimed against those who inveigh against medicine in this manner, confident on account of the kinds of

[1] ἐπιπορ. A: ἐμπορ. M Erotian frag. 36

ψέγει,[2] εὐπορέων δὲ διὰ τὴν τέχνην ᾗ βοηθεῖ, δυνάμενος δὲ διὰ σοφίην ᾗ πεπαίδευται.

2. Δοκέει δή μοι τὸ μὲν σύμπαν τέχνη εἶναι οὐδεμία
4 οὐκ ἐοῦσα· | καὶ γὰρ ἄλογον τῶν ἐόντων τι ἡγεῖσθαι μὴ ἐόν· ἐπεὶ τῶν γε μὴ ἐόντων τίνα ἄν τις οὐσίην θεησάμενος ἀπαγγείλειεν ὡς ἔστιν; εἰ γὰρ δὴ ἔστι γ᾽ ἰδεῖν τὰ μὴ ἐόντα, ὥσπερ τὰ ἐόντα, οὐκ οἶδ᾽ ὅπως ἄν τις αὐτὰ νομίσειε μὴ ἐόντα,[3] ἅ γε εἴη καὶ ὀφθαλμοῖσιν ἰδεῖν καὶ γνώμῃ νοῆσαι ὡς ἔστιν· ἀλλ᾽ ὅπως μὴ οὐκ ᾖ τοῦτο τοιοῦτον· ἀλλὰ τὰ μὲν ἐόντα αἰεὶ ὁρᾶταί τε καὶ γινώσκεται, τὰ δὲ μὴ ἐόντα οὔτε ὁρᾶται οὔτε γινώσκεται. γινώσκεται τοίνυν δεδειγμένων[4] εἴδεα[5] τῶν τεχνέων, καὶ οὐδεμία ἐστὶν ἥ γε ἔκ τινος εἴδεος οὐχ ὁρᾶται. οἶμαι δ᾽ ἔγωγε καὶ τὰ ὀνόματα αὐτὰς διὰ τὰ εἴδεα λαβεῖν· ἄλογον γὰρ ἀπὸ τῶν ὀνομάτων ἡγεῖσθαι τὰ εἴδεα βλαστάνειν, καὶ ἀδύνατον· τὰ μὲν γὰρ ὀνόματα φύσιος νομοθετήματά ἐστι, τὰ δὲ εἴδεα οὐ νομοθετήματα, ἀλλὰ βλαστήματα.

3. Περὶ μὲν οὖν τούτων εἴ γέ τις μὴ ἱκανῶς ἐκ τῶν εἰρημένων ξυνίησιν, ἐν ἄλλοισιν ἂν λόγοισιν σαφέστερον διδαχθείη. περὶ δὲ ἰητρικῆς, ἐς ταύτην γὰρ ὁ λόγος, ταύτης οὖν τὴν ἀπόδειξιν ποιήσομαι. καὶ πρῶτόν γε διοριεῦμαι ὃ νομίζω ἰητρικὴν εἶναι· τὸ δὴ πάμπαν ἀπαλλάσσειν τῶν νοσεόντων τοὺς καμάτους καὶ τῶν νοσημάτων τὰς σφοδρότητας ἀμβλύνειν, καὶ
6 τὸ μὴ ἐγχειρέειν τοῖσι κεκρατημένοις | ὑπὸ τῶν νοση-

[2] οὓς ψέγει M: τοὺς ψέγειν ἐθέλοντας A

people against whom it is directed, supported by the art it defends, and made possible by the insights of its teaching.

2. In principle, I believe that there is no art which does not exist, since it is nonsense to think that any of the things that exist does not exist: for what actual evidence of things that do not exist could anyone have observed, to announce that they exist? Or if in fact it was possible to see the nonexistent in the same way as the existent, I do not know why anyone would think that things were nonexistent that the eyes could see and the mind could grasp as existent? But this could hardly be so. What exists is always seen and known, while what does not exist is never seen or known. Now we know the real essences of the arts that have been revealed, and there is no art that is not seen with some real essence. I myself think that the arts take their names, too, from their essences: for it is not logical to think that essences came from names, and in fact impossible. Names, after all, are just conventions, whereas real essences are not conventions, but the offspring of nature.

3. Now if anyone has not understood sufficiently how it is concerning these matters from what I have said, let him gain a clearer grasp of it from other accounts. As for medicine, which is the subject of this discourse, I will now present my demonstration, beginning with a definition of what I consider medicine to be. In general, medicine is allaying patients' pains and blunting the violence of their diseases, with the proviso of not undertaking cases where people are already overcome by their diseases, knowing

---

[3] Add. ὤσπερ τὰ ἔοντα M [4] δεδειγμένων M: δεδιδαγμένων A [5] εἴδεα Ermerins: ἤδη codd.

μάτων, εἰδότας ὅτι πάντα ταῦτα[6] *δύναται ἰητρική. ὡς οὖν ποιέει τε ταῦτα, καὶ οἵη τέ ἐστι διὰ παντὸς ποιέειν, περὶ τούτου μοι ὁ λοιπὸς λόγος ἤδη ἔσται. ἐν δὲ τῇ τῆς τέχνης ἀποδείξει ἅμα καὶ τοὺς λόγους τῶν αἰσχύνειν αὐτὴν οἰομένων ἀναιρήσω, ᾗ ἂν ἕκαστος αὐτῶν πρήσσειν τι οἰόμενος τυγχάνῃ.*

4. *Ἔστι μὲν οὖν μοι ἀρχὴ τοῦ λόγου ἣ καὶ ὁμολογηθήσεται παρὰ πᾶσιν· ὅτι μὲν ἔνιοι ἐξυγιαίνονται τῶν θεραπευομένων ὑπὸ ἰητρικῆς ὁμολογέεται· ὅτι δ' οὐ πάντες, ἐν τούτῳ ἤδη ψέγεται ἡ τέχνη, καί φασιν οἱ τὰ χείρω λέγοντες διὰ τοὺς ἁλισκομένους ὑπὸ τῶν νοσημάτων τοὺς ἀποφεύγοντας αὐτὰ τύχῃ ἀποφεύγειν καὶ οὐ διὰ τὴν τέχνην. ἐγὼ δὲ οὐκ*[7] *ἀποστερέω μὲν οὐδ' αὐτὸς τὴν τύχην ἔργου οὐδενός, ἡγεῦμαι δὲ τοῖσι μὲν κακῶς θεραπευομένοισι νοσήμασι τὰ πολλὰ τὴν ἀτυχίην ἕπεσθαι, τοῖσι δὲ εὖ τὴν εὐτυχίην. ἔπειτα δὲ καὶ πῶς οἷόν τ'*[8] *ἐστὶ τοῖς ὑγιασθεῖσιν ἄλλο τι αἰτιήσασθαι ἢ τὴν τέχνην, εἴπερ χρώμενοι αὐτῇ καὶ ὑπουργέοντες ὑγιάσθησαν; τὸ μὲν γὰρ τῆς τύχης εἶδος ψιλὸν οὐκ ἐβουλήθησαν θεήσασθαι, ἐν ᾧ τῇ τέχνῃ ἐπέτρεψαν σφᾶς αὐτούς, ὥστε τῆς μὲν ἐς τὴν τύχην ἀναφορῆς ἀπηλλαγμένοι εἰσί, τῆς μέντοι ἐς τὴν τέχνην οὐκ ἀπηλλαγμένοι· ἐν ᾧ γὰρ ἐπέτρεψαν αὐτῇ σφᾶς καὶ ἐπίστευσαν, ἐν τούτῳ αὐτῆς καὶ τὸ εἶδος ἐσκέψαντο καὶ τὴν δύναμιν περανθέντος τοῦ ἔργου ἔγνωσαν.*

5. *Ἐρεῖ δὴ ὁ τἀναντία λέγων,*[9] *ὅτι πολλοὶ ἤδη καὶ*

that medicine is capable of all this. That medicine both does these things and is in all instances capable of doing them will be the subject of the rest of my work. At the same time, in my exposition of the art I will also refute the arguments of those who wish to detract from it, wherever any of them happens to believe he is achieving anything.

4. Now the starting point of my argument is something that will receive agreement from everyone: conceded is that some of those treated by the medical art recover completely. Not all, however, a fact for which the art is already blamed, and those who malign the art argue, on the basis of the patients who die, that the ones who escape do so by chance, and not by means of the art. Now I do not deny that chance has a role, since I believe that among cases where diseases are poorly managed ill fortune usually results, and where they are well managed good fortune. And furthermore, how is it possible for patients who are restored to health to attribute this to anything other than the art, if their cure comes from employing the art and following its instructions? These patients were unwilling to limit their hope for recovery to chance alone, but turned themselves over to the art, thereby freeing themselves from a dependence on pure chance, but not, to be sure, from their dependence on the art. For in turning to the art and putting their trust in it, they recognized its existence, and when it had its effect, they knew its power.

5. Of course my adversary will say that many ill people

---

[6] *π. τ. Med. Def.* (Σ): *τ. π. Med. Def.* (Φ): *ταῦτα* M: *πάντα* A

[7] *οὐκ* M: om. A [8] *οἷόν τ'* M: *οἴονται* A

[9] The text begins here in Urb preceded by the introduction: *ἐροῦσι δέ τινες οἱ τὴν ἰατρικὴν διαβάλλοντες μὴ εἶναι τέχνην*

οὐ χρησάμενοι ἰητρῷ νοσέοντες ὑγιάσθησαν, καὶ ἐγὼ τῷ λόγῳ οὐκ ἀπιστέω· δοκεῖ δέ μοι οἷόν τ'[10] εἶναι καὶ
8 ἰητρῷ μὴ χρωμένους ἰητρικῇ | περιτυχεῖν, οὐ μὴν ὥστε εἰδέναι ὅ τι ὀρθὸν ἐν αὐτῇ ἔνι ἢ ὅ τι μὴ ὀρθόν, ἀλλ' ὥστε ἐπιτύχοιεν τοιαῦτα θεραπεύσαντες ἑωυτούς, ὁποῖάπερ ἂν ἐθεραπεύθησαν εἰ καὶ ἰητροῖσιν ἐχρῶντο. καὶ τοῦτό γε τεκμήριον μέγα τῇ οὐσίῃ τῆς τέχνης, ὅτι ἐοῦσά τέ ἐστι καὶ μεγάλη, ὅπου γε φαίνονται καὶ οἱ μὴ νομίζοντες αὐτὴν εἶναι σῳζόμενοι δι' αὐτήν· πολλὴ γὰρ ἀνάγκη καὶ τοὺς μὴ χρωμένους ἰητροῖσι νοσήσαντας δὲ καὶ ὑγιασθέντας εἰδέναι, ὅτι ἢ δρῶντές τι ἢ μὴ δρῶντες ὑγιάσθησαν· ἢ γὰρ ἀσιτίῃ ἢ πολυφαγίῃ, ἢ ποτῷ πλέονι ἢ δίψῃ, ἢ λουτροῖς, ἢ ἀλουσίῃ, ἢ πόνοισιν ἢ ἡσυχίῃ, ἢ ὕπνοισιν ἢ ἀγρυπνίῃ, ἢ τῇ ἁπάντων τούτων ταραχῇ χρώμενοι ὑγιάσθησαν. καὶ τῷ ὠφελῆσθαι πολλὴ ἀνάγκη αὐτούς ἐστιν ἐγνωκέναι ὅ τι ἦν τὸ ὠφελῆσαν, καὶ εἴ τί γ'[11] ἐβλάβησαν, τῷ βλαβῆναι, ὅ τι ἦν τὸ βλάψαν. τὰ γὰρ τῷ ὠφελῆσθαι καὶ τὰ τῷ βεβλάφθαι ὡρισμένα οὐ πᾶς ἱκανὸς γνῶναι· εἰ τοίνυν ἐπιστήσεται ἢ ἐπαινέειν ἢ ψέγειν ὁ νοσήσας τῶν διαιτημάτων τι οἷσιν ὑγιάνθη, πάντα ταῦτα τῆς ἰητρικῆς ἐστι. καὶ ἔστιν οὐδὲν ἧσσον τὰ ἁμαρτηθέντα τῶν ὠφελησάντων μαρτύρια τῇ τέχνῃ ἐς τὸ εἶναι· τὰ μὲν γὰρ ὠφελήσαντα τῷ ὀρθῶς προσενεχθῆναι ὠφέλησε, τὰ δὲ βλάψαντα τῷ μηκέτι

[10] οἷόν τ' M Urb: οἴονται A
[11] γ' Jouanna: τ' codd.

have recovered who did not make use of a physician, which I do not deny. However, I am convinced that it is possible to stumble upon medical success without the aid of a physician, not such that these patients were aware what is correct in the art or incorrect, but by managing to treat themselves successfully in the same way they would have been treated if they had employed physicians. It is also a compelling piece of evidence for the art's existence, and that it both exists and is powerful, when it is seen that even those denying its existence are saved by it. For it is absolutely certain that people who fail to use a physician, but recover from their disease, must know that they have done so as the result of performing or not performing some thing, since it is by fasting or eating to repletion, taking more drink or going without drink, taking baths or avoiding them, exercising themselves or resting, sleeping or staying awake, or by a mixture of all these, that they have recovered. And from the experience of being benefited, they must have learned what brought the benefits, and when they were harmed, to learn what caused the harm. Not everyone is capable of distinguishing between what is beneficial and what is harmful; but if a sick person does understand how to praise or blame the parts of the dietetic program by which he has recovered, all this belongs to medicine. And mistakes made are no less proofs that the medical art exists, than the benefits it provides: beneficial practices benefit because they are correctly applied, harmful practices harm because they are not cor-

ὀρθῶς προσενεχθῆναι ἔβλαψε. καίτοι ὅπου τό τε ὀρθὸν καὶ τὸ μὴ ὀρθὸν ὅρον ἔχει ἑκάτερον, πῶς τοῦτο οὐκ ἂν τέχνη εἴη; τοῦτο γὰρ ἔγωγέ φημι ἀτεχνίην εἶναι, ὅπου μήτε ὀρθὸν ἔνι μηδὲν μήτε οὐκ ὀρθόν· ὅπου δὲ τούτων ἔνεστιν ἑκάτερον, οὐκέτι ἂν τοῦτο τὸ ἔργον ἀτεχνίης εἴη.[12]

6. Ἔτι τοίνυν εἰ μὲν ὑπὸ φαρμάκων τῶν τε καθαιρόντων καὶ | τῶν ἱστάντων ἡ ἴησις τῇ τε ἰητρικῇ καὶ τοῖσιν ἰητροῖσι μοῦνον ἐγίνετο, ἀσθενὴς ἦν ἂν ὁ ἐμὸς λόγος· νῦν δὲ φαίνονται τῶν ἰητρῶν οἱ μάλιστα ἐπαινεόμενοι καὶ διαιτήμασιν ἰώμενοι καὶ ἄλλοισί γε εἴδεσιν, ἃ οὐκ ἄν τις φαίη, μὴ ὅτι ἰητρός, ἀλλ᾽ οὐδὲ ἰδιώτης ἀνεπιστήμων ἀκούσας, μὴ οὐ τῆς τέχνης εἶναι. ὅπου οὖν οὐδὲν οὔτ᾽ ἐν τοῖς ἀγαθοῖσι τῶν ἰητρῶν οὔτ᾽ ἐν τῇ ἰητρικῇ αὐτῇ ἀχρεῖόν ἐστιν, ἀλλ᾽ ἐν τοῖσι πλείστοισι τῶν τε φυομένων καὶ τῶν ποιευμένων ἔνεστι τὰ εἴδεα τῶν θεραπειῶν καὶ τῶν φαρμάκων, οὐκ ἔστιν ἔτι οὐδενὶ τῶν ἄνευ ἰητροῦ ὑγιαζομένων τὸ αὐτόματον αἰτιήσασθαι ὀρθῷ λόγῳ· τὸ μὲν γὰρ αὐτόματον οὐδὲν φαίνεται ἐὸν ἐλεγχόμενον· πᾶν γὰρ τὸ γινόμενον διά τι εὑρίσκοιτ᾽ ἂν γινόμενον, καὶ ἐν τῷ διά τι τὸ αὐτόματον οὐ φαίνεται οὐσίην ἔχον οὐδεμίην ἀλλ᾽ ἢ ὄνομα· ἡ δὲ ἰητρικὴ καὶ ἐν τοῖσι διά τι καὶ ἐν τοῖσι προνοουμένοισι φαίνεταί τε καὶ φανεῖται αἰεὶ[13] οὐσίην ἔχουσα.

7. Τοῖσι μὲν οὖν τῇ τύχῃ τὴν ὑγιείην προστιθεῖσι τὴν δὲ τέχνην ἀφαιρέουσι τοιαῦτ᾽ ἄν τις λέγοι· τοὺς δ᾽ ἐν τῇσι τῶν ἀποθνῃσκόντων ξυμφορῇσι τὴν τέ-

rectly applied. For indeed when there is a difference between what is correct and what incorrect, how could this not represent an art? Here in fact is what I would define as the lack of an art: where there is no idea of right or wrong: when, however, these two exist, such an activity can no longer be said to lack an art.

6. Now furthermore, if the treatments of medicine and physicians consisted only in medications that clean and contract, my case would be weak. But as it is, those physicians who are most praised appear to cure by dietetic methods and other forms (sc. of treatment), which no one, on hearing mentioned—either physician or untutored layman—would deny belong to the art. For when among good physicians and in the medical art itself, there is almost nothing that lacks a use, but among most things that grow or are made there are healing measures and medications, it is no longer possible, in the case of someone healed without a physician, rationally to find only some spontaneous cause operative. When the matter is carefully weighed, no spontaneous cause is seen to exist, for everything that happens is discovered to happen for some reason, and with this "some reason" it becomes clear that spontaneity has no real existence, besides being a word. Medicine, however, both in its causality and its predictability is seen, and will always be seen, to possess reality.

7. The following might be said against those who attribute recovery to chance and deny that the art exists. I am also surprised how those who, from the misfortunes of

---

[12] The text in Urb ends here.

[13] αἰεὶ A: ἔτι M

χνην ἀφανίζοντας θαυμάζω, ὁτέῳ ἐπαιρόμενοι ἀξιοχρέῳ λόγῳ τὴν μὲν τῶν ἀποθνῃσκόντων ἀκρασίην ἀναιτίαν[14] καθιστᾶσι, τὴν δὲ τῶν τὴν ἰητρικὴν μελετησάντων ξύνεσιν αἰτίην· ὡς τοῖσι μὲν ἰητροῖς ἔνεστι τὰ μὴ δέοντα ἐπιτάξαι, τοῖσι δὲ νοσέουσιν οὐκ ἔνεστι τὰ προσταχθέντα παραβῆναι. καὶ μὴν πολύ γε εὐλογώτερον τοῖσι κάμνουσιν ἀδυνατέειν τὰ προστασσόμενα ὑπουργέειν, ἢ τοῖς ἰητροῖσι τὰ μὴ δέοντα ἐπιτάσσειν. οἱ μὲν γὰρ ὑγιαινούσῃ γνώμῃ μεθ᾽ ὑγιαίνοντος σώματος ἐγχειρέουσι, λογισάμενοι τά τε παρεόντα, τῶν τε παροιχομένων τὰ ὁμοίως διατεθέντα τοῖσι παρεοῦσιν, ὥστε ποτὲ θεραπευθέντα εἰπεῖν ὡς

12 ἀπήλλαξαν· οἱ δὲ οὔτε ἃ κάμνουσιν | οὔτε δι᾽ ἃ κάμνουσιν,[15] οὐδ᾽ ὅ τι ἐκ τῶν παρεόντων ἔσται, οὐδ᾽ ὅ τι ἐκ τῶν τούτοισιν ὁμοίων γίνεται εἰδότες, ἐπιτάσσονται, ἀλγέοντες μὲν ἐν[16] τῷ παρεόντι, φοβεύμενοι δὲ τὸ μέλλον, καὶ πλήρεες μὲν τῆς νούσου, κενεοὶ δὲ σιτίων, ἐθέλοντες δὲ τὰ πρὸς τὴν νοῦσον ἤδη μᾶλλον ἢ τὰ πρὸς τὴν ὑγιείην προσδέχεσθαι, οὐκ ἀποθανεῖν ἐρῶντες ἀλλὰ καρτερεῖν ἀδυνατέοντες. οὕτω δὲ διακειμένους πότερον εἰκὸς τούτους τὰ ὑπὸ τῶν ἰητρῶν ἐπιτασσόμενα ποιέειν ἢ ἄλλα ποιέειν ἢ ἃ ἐπετάχθησαν, ἢ τοὺς ἰητροὺς τοὺς ἐκείνως διακειμένους ὡς ὁ πρόσθεν λόγος ἡρμήνευσεν ἐπιτάσσειν τὰ μὴ δέοντα; ἆρ᾽ οὐ πολὺ μᾶλλον, τοὺς μὲν δεόντως ἐπιτάσσειν τοὺς δὲ εἰκότως ἀδυνατέειν πείθεσθαι, μὴ πειθομένους δὲ

[14] ἀκρασίην αἰτίη M: ἀτυχίην ἀναιτίαν A

patients who succumb, deny that the art has any existence, are then induced, by I know not what adequate reasoning, to ignore that the misfortune of the dying patients could be caused by their own lack of discipline, but instead make the knowledge of the practitioners of the medical art the cause—as if physicians could give wrong instructions, but it was not possible for patients to disobey the instructions. In fact, it is much more probable that patients will be incapable of following instructions, than that physicians will give wrong instructions: the latter, physicians healthy in mind and body, are undertaking their treatment by judging the events of the present and those of the past likely to have the same effects again, so that they are capable of stating what treatment once brought alleviation, while the former, patients who do not understand what they are suffering or why, nor what will develop out of the present or out of circumstances similar to these, are receiving their instructions while subject to pain in the present, and fearing for the future, full of the disease, empty of food, desirous more of receiving some immediate remedy against their disease than of regaining their health, wanting not to die, but being incapable of endurance. With patients in this condition, is it more likely for them to ignore their physicians' orders and do something other than they are instructed, or for physicians, who are such as I have indicated in my account above, to give incorrect instructions? Is it not much more probable that the physician will give correct instructions, but that the patient will be unable to obey these, and from this disobedience will meet his

---

[15] εἰδότες add. A [16] ἐν A: om. M

περιπίπτειν τοῖσι θανάτοις, ὧν οἱ μὴ ὀρθῶς λογιζόμενοι τὰς αἰτίας τοῖς οὐδὲν αἰτίοις ἀνατιθέασι, τοὺς αἰτίους ἐλευθεροῦντες;

8. Εἰσὶ δέ τινες οἳ καὶ διὰ τοὺς μὴ θέλοντας ἐγχειρέειν τοῖσι κεκρατημένοις ὑπὸ τῶν νοσημάτων μέμφονται τὴν ἰητρικήν, λέγοντες ὡς ταῦτα μὲν καὶ αὐτὰ ὑφ᾽ ἑαυτῶν ἂν ἐξυγιάζοιτο ἃ ἐγχειρέουσιν ἰᾶσθαι, ἃ δ᾽ ἐπικουρίης δεῖται μεγάλης[17] οὐχ ἅπτονται, δεῖν δέ, εἴπερ ἦν ἡ τέχνη, πάνθ᾽ ὁμοίως ἰᾶσθαι. οἱ μὲν οὖν ταῦτα λέγοντες, εἰ ἐμέμφοντο τοῖς ἰητροῖς, ὅτι αὐτῶν τοιαῦτα λεγόντων οὐκ ἐπιμέλονται ὡς παραφρονεύντων, εἰκότως ἂν ἐμέμφοντο μᾶλλον ἢ ἐκεῖνα μεμφόμενοι. εἰ γάρ τις ἢ τέχνην ἐς ἃ μὴ τέχνη, ἢ φύσιν ἐς ἃ μὴ φύσις[18] πέφυκεν, ἀξιώσειε δύνασθαι, ἀγνοεῖ

14 ἄγνοιαν ἁρμόζουσαν μανίῃ[19] μᾶλλον | ἢ ἀμαθίῃ. ὧν γὰρ ἔστιν ἡμῖν τοῖσί τε τῶν φυσίων τοῖσι τε τῶν τεχνέων ὀργάνοις ἐπικρατέειν, τούτων ἔστιν ἡμῖν δημιουργοῖς εἶναι, ἄλλων δὲ οὐκ ἔστιν. ὅταν οὖν τι πάθῃ ὥνθρωπος κακὸν ὃ κρέσσον ἐστὶ τῶν ἐν ἰητρικῇ ὀργάνων, οὐδὲ προσδοκᾶσθαι τοῦτό που δεῖ ὑπὸ ἰητρικῆς κρατηθῆναι ἄν. αὐτίκα γὰρ τῶν ἐν ἰητρικῇ καιόντων τὸ πῦρ ἐσχάτως καίει, τούτου δὲ ἡσσόνως ἄλλα πολλά· τῶν μὲν οὖν ἡσσόνων τὰ κρέσσω οὔπω δῆλον ὅτι ἀνίητα· τῶν δὲ κρατίστων τὰ κρέσσω πῶς

[17] μεγάλης A: om. M [18] φύσις A: om. M

[19] ἄγνοιαν . . . μανίῃ Diels: ἄγνοιαν . . . μανίη Zwinger in marg.: ἄγνοιη . . . μανείην M: ἄγνοια . . . μανίην A

death, whose causes false reasoners will ascribe to those who are not to blame, while letting those truly to blame off scot-free?

8. There are also people who censure medicine because of practitioners who are unwilling to take on patients who are overwhelmed by their diseases, saying that these are willing to treat diseases that would heal completely by themselves, but refuse to touch cases which require great help, but that if there were a medical art, it should treat all cases equally. Now if the critics arguing like this censured physicians who refuse to treat people who make arguments like the ones they themselves are making, on the grounds that they are mad, they would be bringing a more plausible charge than the one they do bring. For anyone expecting from an art a power over what does not belong to the art, or from nature a power over what does not belong to nature, would be erring with an ignorance more akin to madness than to lack of knowledge. For in areas where we are able to gain the mastery by means of their nature or the tools of the art, it is possible for us to be craftsmen, but not in others. Thus when a patient suffers some evil more powerful than the means of the medical art, it cannot be expected that this should be in some way overcome. For example, of the caustics used in medicine fire is the most powerful, while there are many others which are less powerful than it: conditions stronger than these less powerful caustics are thus clearly not yet irremediable, but conditions more powerful than the stron-

οὐ δῆλον ὅτι ἀνίητα; ἃ γὰρ πῦρ δημιουργέει, πῶς οὐ τὰ τούτῳ μὴ ἁλισκόμενα δηλοῖ ὅτι ἄλλης τέχνης δεῖται, καὶ οὐ ταύτης, ἐν ᾗ τὸ πῦρ ὄργανον; ὡὑτὸς δέ μοι λόγος καὶ ὑπὲρ τῶν ἄλλων ὅσα τῇ ἰητρικῇ ξυνεργεῖ, ὧν ἁπάντων φημὶ δεῖν ἑκάστου <μὴ>[20] κατατυχόντα τὸν ἰητρὸν τὴν δύναμιν αἰτιᾶσθαι τοῦ πάθεος, ἀλλὰ μὴ τὴν τέχνην. οἱ μὲν οὖν μεμφόμενοι τοὺς τοῖσι κεκρατημένοισι μὴ ἐγχειρέοντας παρακελεύονται καὶ ὧν μὴ προσήκει ἅπτεσθαι οὐδὲν ἧσσον ἢ ὧν προσήκει· παρακελευόμενοι δὲ ταῦτα ὑπὸ μὲν τῶν ὀνόματι ἰητρῶν θαυμάζονται, ὑπὸ δὲ τῶν καὶ τέχνῃ καταγελῶνται. οὐ μὴν οὕτως ἀφρόνων οἱ ταύτης τῆς δημιουργίης ἔμπειροι οὔτε μωμητῶν οὔτ᾽ αἰνετῶν δέονται, ἀλλὰ λελογισμένων πρὸς ὅ τι αἱ ἐργασίαι τῶν δημιουργῶν τελευτώμεναι πλήρεις εἰσί, καὶ ὅτευ ὑπολειπόμεναι ἐνδεεῖς, ἔτι τε τῶν ἐνδειῶν, ἅς τε τοῖς δημιουργοῦσιν ἀναθετέον ἅς τε τοῖς δημιουργεομένοισι. |

16 9. Τὰ μὲν οὖν κατὰ τὰς ἄλλας τέχνας ἄλλος χρόνος μετ᾽ ἄλλου λόγου δείξει· τὰ δὲ κατὰ τὴν ἰητρικὴν οἷά τέ ἐστιν ὥς τε κριτέα, τὰ μὲν ὁ παροιχόμενος τὰ δὲ ὁ παρεὼν διδάξει λόγος. ἔστι γὰρ τοῖσι ταύτην τὴν τέχνην ἱκανῶς εἰδόσι τὰ μὲν τῶν νοσημάτων οὐκ ἐν δυσόπτῳ κείμενα, καὶ οὐ πολλά, τὰ δὲ οὐκ ἐν εὐδήλῳ καὶ πολλά. ἔστι δὲ τὰ μὲν ἐξανθεῦντα ἐς τὴν χροιὴν ἢ χροιῇ[21] ἢ οἰδήμασιν ἐν εὐδήλῳ· παρέχει γὰρ ἑωυ-

[20] μὴ add. recc. Littré [21] ἢ χροιῇ A: om. M

gest caustics, how could they not be clearly irremediable? But how does a condition which is not overcome by what fire can achieve fail to demonstrate that it requires some art other than the one in which fire is the most potent means? My argument is the same concerning the other helpful agents which play a role in medicine: I claim that when a physician fails when using any of these, he must blame the strength of the disease, and not the art. Now those blaming physicians for not taking on desperate cases urge them to treat unsuitable patients no less than suitable ones. In recommending such things they may be admired by physicians in name only, but will be scorned by those who (sc. besides the name) also have a command of the art. But practitioners experienced in this art have need of neither the blame nor the praise of such fools, but of judges with an understanding of how far the techniques of the experts are fully perfected, and in how far they are wanting, and have considered which of their failures are to be attributed to the practitioners, and which to those who are being treated.

9. Now how things are in the other arts will be explained on another occasion in another discourse. Regarding medicine, its essence and how it is to be judged, the previous and present argument will show. Now facing practitioners adequately knowledgeable in this art are both diseases whose site is not difficult to observe—which are not many—and diseases whose site is not easily discerned—which are many. The first are lesions on the skin whose site is easily observable by both their color and their

τῶν τῇ τε ὄψει τῷ τε ψαῦσαι τὴν στερεότητα καὶ τὴν ὑγρότητα αἰσθάνεσθαι, καὶ ἅ τε αὐτῶν θερμὰ ἅ τε ψυχρά, ὧν τε ἑκάστου ἢ παρουσίῃ ἢ ἀπουσίῃ[22] τοιαῦτ' ἐστιν. τῶν μὲν οὖν τοιούτων πάντων ἐν πᾶσι τὰς ἀκέσιας ἀναμαρτήτους δεῖ εἶναι, οὐχ ὡς ῥηϊδίας, ἀλλ' ὅτι ἐξεύρηνται· ἐξεύρηνταί γε μὴν οὐ τοῖσι βουληθεῖσιν, ἀλλὰ τούτων τοῖσι δυνηθεῖσιν· δύνανται δὲ οἷσι τά τε τῆς παιδείης μὴ ἐκποδὼν τά τε τῆς φύσιος μὴ ἀταλαίπωρα.[23]

10. Πρὸς μὲν οὖν τὰ φανερὰ τῶν νοσημάτων οὕτω δεῖ εὐπορέειν τὴν τέχνην· δεῖ γε μὴν αὐτὴν οὐδὲ πρὸς τὰ ἧσσον φανερὰ ἀπορέειν· ἔστι δὲ ταῦτα ἃ πρός τε τὰ ὀστέα τέτραπται καὶ τὴν νηδύν· ἔχει δὲ τὸ σῶμα οὐ μίαν, ἀλλὰ πλείους· δύο μὲν γὰρ αἱ τὸ σιτίον δεχόμεναί τε καὶ ἀφιεῖσαι, ἄλλαι δὲ τούτων πλείους, ἃς ἴσασιν οἷσι τούτων ἐμέλησεν· ὅσα γὰρ τῶν μελέων ἔχει σάρκα περιφερέα, ἣν μῦν καλέουσιν, πάντα νηδὺν ἔχει. πᾶν γὰρ τὸ ἀσύμφυτον, ἤν τε δέρματι, ἤν τε σαρκὶ καλύπτηται, κοῖλόν ἐστιν· πληροῦταί τε ὑγιαῖνον μὲν πνεύματος ἀσθενῆσαν δὲ ἰχῶρος· ἔχουσι 18 μὲν τοίνυν οἱ βραχίονες σάρκα | τοιαύτην· ἔχουσι δ' οἱ μηροί· ἔχουσι δ' αἱ κνῆμαι. ἔτι δὲ καὶ ἐν τοῖσιν ἀσάρκοισι τοιαύτη ἔνεστιν οἵη καὶ ἐν τοῖσιν εὐσάρκοισιν ἐνεῖναι δέδεκται·[24] ὅ τε γὰρ θώρηξ καλεόμενος, ἐν ᾧ τὸ ἧπαρ στεγάζεται, ὅ τε τῆς κεφαλῆς κύκλος, ἐν ᾧ ὁ ἐγκέφαλος, τό τε νῶτον, πρὸς ᾧ ὁ πλεύμων,

[22] παρουσίῃ ἢ ἀπουσίῃ Littré: παρουσίη ἢ ἀπουσίη codd.

swellings; it is possible to ascertain their solidity or moistness by sight or touch, and know which of them are warm or cold, and which signs, by their presence or absence, determine the diseases to be such as they are. Now the treatments of all such disorders should always be infallible, not because they are easy, but because they have been fully discovered; their discovery has been made not by people who merely have such a desire, but by those who have power over them, the power possessed by practitioners whose education is not faulty, and whose nature is not indolent.

10. Against such visible diseases, then, this is why the art must be successful; but against less visible ones, too, it should not be at a loss. These are the conditions which turn toward the bones and the belly. Here the body has not one, but many cavities, two which respectively receive the food and discharge it, and others in addition, known to those who are interested in such things. All the limbs, for example, that have flesh around them which is called muscle, contain a cavity—for everything that is not continuous, whether covered by skin or flesh, is hollow—which is filled when healthy with breath, but when ill with ichor. For example, the arms have flesh like this, as do the thighs and lower legs, and even in parts without flesh the same has been demonstrated to be present as in the fleshy parts. And what is called the thorax (which shelters the liver), and the circle of the head (which contains the brain), and the back (against which the lung lies): none of

23 ἀταλ. A: ταλ. M 24 δέδεκται M: λέλεκται A

τούτων οὐδὲν ὅ τι οὐ καὶ αὐτὸ κενόν ἐστι, πολλῶν διαφυσίων μεστόν· ἔστι δ' οἷσιν οὐδὲν ἀπέχει πολλῶν ἀγγεῖα εἶναι, τῶν μέν τι βλαπτόντων τὸν κεκτημένον, τῶν δὲ καὶ ὠφελεύντων. ἔτι δὲ καὶ πρὸς τούτοισι φλέβες πολλαὶ καὶ νεῦρα οὐκ ἐν τῇ σαρκὶ μετέωρα, ἀλλὰ πρὸς τοῖς ὀστέοισι προστεταμένα, ‹ἃ›[25] σύνδεσμός ἐστι τῶν ἄρθρων, καὶ αὐτὰ τὰ ἄρθρα, ἐν οἷσιν αἱ συμβολαὶ τῶν κινεομένων ὀστέων ἐγκυκλέονται· καὶ τούτων οὐδὲν ὅ τι οὐχ ὕπαφρόν ἐστι καὶ ἔχον περὶ αὐτὸ θαλάμας, ἃς καταγγέλλει ἰχώρ, ὃς ἐκδιοιγομένων αὐτῶν πολύς τε καὶ πολλὰ λυπήσας ἐξέρχεται.

11. Οὐ γὰρ δὴ ὀφθαλμοῖσί γε ἰδόντι τούτων τῶν εἰρημένων οὐδενὶ οὐδὲν ἔστιν εἰδέναι· διὸ καὶ ἄδηλα ἐμοί τε ὠνόμασται καὶ τῇ τέχνῃ κέκριται εἶναι. οὐ μὴν ὅτι ἄδηλα κεκράτηκεν, ἀλλ' ᾗ δυνατὸν κεκράτηται· δυνατὸν δέ ὥς αἵ τε τῶν νοσεόντων φύσιες ἐς τὸ σκεφθῆναι παρέχουσιν, αἵ τε τῶν ἐρευνησόντων ἐς 20 τὴν ἔρευναν πεφύκασιν. | μετὰ πλείονος μὲν γὰρ πόνου καὶ οὐ μετ' ἐλάσσονος χρόνου ἢ εἰ τοῖσιν ὀφθαλμοῖσιν ἑωρᾶτο γινώσκεται· ὅσα γὰρ τὴν τῶν ὀμμάτων ὄψιν ἐκφεύγει, ταῦτα τῇ τῆς γνώμης ὄψει κεκράτηται· καὶ ὅσα δὲ ἐν τῷ μὴ ταχὺ ὀφθῆναι οἱ νοσέοντες πάσχουσιν, οὐχ οἱ θεραπεύοντες αὐτοὺς αἴτιοι, ἀλλ' ἡ φύσις ἥ τε τοῦ νοσέοντος ἥ τε τοῦ νοσήματος· ὁ μὲν γάρ, ἐπεὶ οὐκ ἦν αὐτῷ ὄψει ἰδεῖν τὸ μοχθέον οὐδ' ἀκοῇ πυθέσθαι, λογισμῷ μετῄει. καὶ γὰρ δὴ καὶ ἃ

[25] Add. Diels

these lacks a hollow, which is filled with many interstices. Some of these are not prevented from being reservoirs filled with many (sc. liquids), some of which bring harm to their possessor, and others which benefit him. Furthermore, in addition to these there are many vessels and cords, located not toward the surface of the flesh, but stretched tightly against the bones, including the ligaments of the joints and the joints themselves—in which the facets of the moving bones rotate. All these parts, without exception, are frothy and contain chambers around them, which the ichor reveals when they open up, as it is expelled in a great amount and provokes violent pain.

11. Now I admit that no person, at least by seeing with his eyes, would learn any of the things being explained here. This is why I call the diseases occult, and the art judges them the same way. Though they are occult, however, such diseases need not be victorious, but as far as it is possible they are conquered—possible to the degree that the natures of patients can be investigated, and that the natures of the investigators include a gift for investigation. For their discovery requires more effort and no less time than do diseases that can be seen with the eyes: this is because what escapes the vision of the eyes must be conquered by the vision of the mind. And what patients suffer from this not being immediately perceived is caused not by the healers, but by the nature of the patient and of the disease. For when the practitioner's vision was not able to see the suffering part or detect it by sound, he pursued it by reasoning. What patients with obscure conditions

πειρῶνται οἱ τὰ ἀφανέα νοσέοντες ἀπαγγέλλειν περὶ τῶν νοσημάτων τοῖσι θεραπεύουσιν, δοξάζοντες μᾶλλον ἢ εἰδότες ἀπαγγέλλουσιν· εἰ γὰρ ἠπίσταντο, οὐκ ἂν περιέπιπτον αὐτοῖσι· τῆς γὰρ αὐτῆς ξυνέσιός ἐστιν ἧσπερ τὸ εἰδέναι τῶν νούσων τὰ αἴτια καὶ τὸ θεραπεύειν αὐτὰς ἐπίστασθαι πάσῃσι τῇσι θεραπείῃσιν αἳ κωλύουσι τὰ νοσήματα μεγαλύνεσθαι. ὅτε οὖν οὐδὲ ἐκ τῶν ἀπαγγελλομένων ἔστι τὴν ἀναμάρτητον σαφήνειαν ἀκοῦσαι, προσοπτέον τι καὶ ἄλλο τῷ θεραπεύοντι·[26] ταύτης οὖν τῆς βραδυτῆτος οὐχ ἡ τέχνη, ἀλλ' ἡ φύσις αἰτίη ἡ τῶν σωμάτων· ἡ μὲν γὰρ αἰσθανομένη ἀξιοῖ θεραπεύειν, καὶ[27] σκοποῦσα ὅπως μὴ τόλμῃ μᾶλλον ἢ γνώμῃ, καὶ ῥᾳστώνῃ μᾶλλον ἢ βίῃ θεραπεύῃ· ἡ δ' ἢν μὲν διεξαρκέσῃ ἐς τὸ ὀφθῆναι, ἐξαρκέσει καὶ ἐς τὸ ὑγιανθῆναι· ἢν δ' ἐν ᾧ τοῦτο ὁρᾶται κρατηθῇ διὰ τὸ βραδέως αὐτὸν ἐπὶ τὸν θεραπεύσοντα ἐλθεῖν ἢ διὰ τὸ τοῦ νοσήματος τάχος, οἰχήσεται. ἐξ ἴσου μὲν γὰρ ὁρμώμενον τῇ θεραπείῃ οὐκ ἔστι θᾶσσον, προλαβὸν δὲ θᾶσσον· προλαμβάνει δὲ διά τε τὴν τῶν σωμάτων στεγνότητα,[28] ἐν ᾗ οὐκ ἐν εὐόπτῳ οἰκέουσιν αἱ νοῦσοι, διά τε τὴν τῶν | καμνόντων ὀλιγωρίην <ἣ>[29] ἐπιτίθεται· οὐ λαμβανόμενοι γάρ, ἀλλ' εἰλημμένοι ὑπὸ τῶν νοσημάτων ἐθέλουσι θεραπεύεσθαι.

22

12. Ἐπεὶ τῆς γε τέχνης τὴν δύναμιν ὁπόταν τινὰ τῶν τὰ ἄδηλα νοσεύντων ἀναστήσῃ, θαυμάζειν ἀξι-

[26] καὶ ἄλλο . . . θεραπεύοντι om. A

attempt themselves to report about their diseases to their attendants is more what they imagine than what they actually know; for if they had understood their diseases, they would not have fallen into them. This follows from the fact that the same understanding which reveals the causes of diseases also provides an understanding of all their treatments, which prevents them from increasing. Thus, when a report that is heard fails to give an unerring clarity, the healer is forced to search for something else. This delay is be blamed not on the art, but on the nature of bodies. Once the disease has been understood, the art takes it upon itself to treat, and is careful to act not with daring, but with knowledge, and it treats with mildness much more than with force: if the body can hold out until it (sc. the cause of the disease) is seen, the disease will hold out until it is healed; but if during the time of this "being seen" the patient is overcome due to the slowness of his going (sc. to the physician), or outrun by the speed of the disease, he will die; for if the disease starts evenly with the therapy, it will not be the winner, but if it has a head start it will. This head start is due to the density of the bodies in which the disease lurks invisible, and derives in addition from the patients' carelessness, since they are not willing to be treated while the disease is taking hold of them, but only when it has already taken hold.

12. When the power of the art cures a patient suffering from one of the occult diseases, this is more remarkable

---

[27] *καὶ* A: om. M
[28] *στεγν.* M: *στεν.* A
[29] Add. Jouanna

ώτερον, ἢ ὁπόταν μὴ ἐγχειρήσῃ τοῖς ἀδυνάτοις. οὔκουν ἐν ἄλλῃ γε δημιουργίῃ τῶν ἤδη εὑρημένων οὐδεμιῇ ἔνεστιν οὐδὲν τοιοῦτον. ἀλλ' αὐτῶν ὅσαι πυρὶ δημιουργεῦνται, τούτου μὴ παρεόντος ἀεργοί εἰσι· καὶ ὅσαι μετὰ τοῦ ὀφθῆναι ἐνεργοὶ καὶ τοῖσιν εὐεπανορθώτοισι σώμασι δημιουργεῦνται, αἱ μὲν μετὰ ξύλων, αἱ δὲ μετὰ σκυτέων, αἱ δὲ [γραφῇ][30] χαλκῷ τε καὶ σιδήρῳ καὶ τοῖσι τούτων ὁμοίοις χύμασιν[31] αἱ πλεῖσται, τὰ δὲ ἐκ τούτων καὶ μετὰ τούτων δημιουργεύμενα, καὶ εὐεπανόρθωτα, ὅμως οὐ τῷ τάχει μᾶλλον ἢ ὡς δεῖ δημιουργεῖται· οὐδ' ὑπερβατῶς, ἀλλ' ἢν ἀπῇ τι τῶν ὀργάνων, ἐλινύει· καίτοι κἀκείναις τὸ βραδὺ πρὸς τὸ λυσιτελεῦν ἀσύμφορον, ἀλλ' ὅμως προτιμᾶται.

13. (12 L.) Ἰητρικὴ δὲ τοῦτο μὲν τῶν ἐμπύων τοῦτο δὲ τῶν τὸ ἧπαρ ἢ τοὺς νεφροὺς τοῦτο δὲ τῶν ξυμπάντων τῶν ἐν τῇ νηδύι νοσεύντων | ἀπεστερημένη τι ἰδεῖν ὄψει ᾗ τὰ πάντα πάντες ἱκανωτάτως ὁρῶσιν, ὅμως ἄλλας εὐπορίας συνεργοὺς εὗρε. φωνῆς τε γὰρ λαμπρότητι καὶ τρηχύτητι, καὶ πνεύματος ταχυτῆτι καὶ βραδυτῆτι, καὶ ῥευμάτων ἃ διαρρεῖν εἴωθεν ἑκάστοισι δι' ὧν ἔξοδοι δέδονται [ὧν][32] τὰ μὲν ὀδμῇσι τὰ δὲ χροίῃσι τὰ δὲ λεπτότητι καὶ παχύτητι διασταθμωμένη τεκμαίρεται ὧν τε σημεῖα ταῦτα, ἅ τε πεπονθότων ἅ τε παθεῖν δυναμένων. ὅταν δὲ ταῦτα τὰ[33] μηνύοντα μηδ' αὐτὴ ἡ φύσις ἑκοῦσα ἀφίῃ, ἀνάγκας

24

[30] Del. Schwartz in Diels

than that medicine refuses to take on someone impossible to heal. For after all, in none of the other crafts that have so far been discovered are there any such situations. In the crafts that are performed with fire, when fire is not present, they do not proceed; in the arts operative in the realm of sight, that practice on easily formed materials, some with woods, others with leathers, and other more numerous ones with bronze, iron, and ingots of other similar things—in all these arts, although repairs are easy, the objects being crafted from these or with these are not made faster than they should be, but according to correctness, and not skipping a step, but if one of the necessary tools is absent, the process stops: so even in these arts, in spite of the fact that slowness is inconvenient for gain, it is still preferred.

13. Where medicine is prevented, e.g., in cases of internal suppuration or conditions of the liver or kidneys, or all those involving the lower abdomen, from seeing anything by means of the sense of sight, by which everyone observes all things most effectively, it has discovered other auxiliary methods (sc. of examination). Thus signs such as the clarity or hoarseness of the voice, the fastness or slowness of breathing, the characteristics of each of the fluxes that routinely pass through various passageways, such as their odors or colors, or their thinness or thickness, are weighed in determining which parts this evidence pertains to, and what these have suffered and what they can suffer. Even when nature does not surrender this information

[31] ὁμ. χ. Heiberg: ὁμοιοχύμασι A: ὁμοίοισιν M
[32] Del. Ermerins [33] τὰ A: om. M

εὕρηκεν ᾗσιν ἡ φύσις ἀζήμιος βιασθεῖσα μεθίησιν· ἀνεθεῖσα δὲ δηλοῖ τοῖσι τὰ τῆς τέχνης εἰδόσιν ἃ ποιητέα. βιάζεται δὲ τοῦτο μὲν πύου τὸ σύντροφον φλέγμα διαχεῖν σιτίων δριμύτητι καὶ πωμάτων, ὅπως τεκμαρεῖταί τι ὀφθὲν περὶ ἐκείνων ὧν αὐτῇ ἐν ἀμηχάνῳ τὸ ὀφθῆναι ἦν· τό τ᾽ αὖ πνεῦμα ὧν κατήγορον ὁδοῖσί τε προσάντεσι καὶ δρόμοις ἐκβιᾶται κατηγορέειν· ἱδρῶτάς τε τούτοισι τοῖσι προειρημένοις ἄγουσα, <καὶ>[34] *ὑδάτων θερμῶν ἀποπνοίῃσι*[35] *τεκμαίρεται. ἔστι δὲ ἃ καὶ διὰ τῆς κύστιος διελθόντα ἱκανώτερα δηλῶσαι τὴν νοῦσόν ἐστιν ἢ διὰ τῆς σαρκὸς ἐξιόντα. ἐξεύρηκεν οὖν καὶ τοιαῦτα πώματα καὶ βρώματα, ἃ τῶν θερμαινόντων θερμότερα γινόμενα* 26 *τήκει τε ἐκεῖνα καὶ* | *διαρρεῖν ποιέει, ᾗ*[36] *οὐκ ἂν διερρύη μὴ*[37] *τοῦτο παθόντα. ἕτερα μὲν οὖν πρὸς ἑτέρων, καὶ ἄλλα δι᾽ ἄλλων ἐστὶ τά τε διιόντα τά τ᾽ ἐξαγγέλλοντα, ὥστ᾽ οὐ θαυμάσιον αὐτῶν τάς τε πιστίας χρονιωτέρας γίνεσθαι τάς τ᾽ ἐγχειρήσιας βραχυτέρας, οὕτω δι᾽ ἀλλοτρίων ἑρμηνειῶν πρὸς τὴν θεραπεύουσαν σύνεσιν ἑρμηνευομένων.*

14. (13 L) *Ὅτι μὲν οὖν καὶ λόγους ἐν ἑωυτῇ εὐπόρους ἐς τὰς ἐπικουρίας ἔχει ἡ ἰητρική, καὶ οὐκ εὐδιορθώτοισι δικαίως οὐκ ἂν ἐγχειρέοι τῇσι νούσοισιν, ἢ ἐγχειρευμένας ἀναμαρτήτους ἂν παρέχοι, οἵ τε νῦν λεγόμενοι λόγοι δηλοῦσιν αἵ τε τῶν εἰδότων τὴν τέ-*

[34] Jouanna
[35] Add. *πυρὶ ὅσα* A

willingly, the art has developed methods of compelling nature to surrender it without any damage, thereby showing those knowledgeable in the art what should be done. For example, the art compels the innate humor phlegm by means of sharp foods and drinks to expel pus, so that some visible evidence is revealed which the art did not previously have the means of seeing. Then again breath is forced through opposing paths and courses to reveal the things it is capable of revealing. By inducing sweats in the ways described above, and by exhalations from hot water, the art also provides evidence. There are also things passing through the bladder which are more important for revealing a disease, than if passing through the flesh. Medicine has also discovered such drinks and foods that, becoming hotter than fomentations, melt things and make them flow out where they would not have flowed, if they had not been compelled. Now as the relationship between the excretions and what they signify is variable and depends on different factors, it is no wonder that trustworthy evidence from them requires greater time, and their treatments are shorter, since it is extraneous criteria that are transmitting knowledge valuable for the understanding of therapy.

14. Thus, that medicine contains within itself reasoning effective in bringing benefit, that it has a reason not to take on cases unamenable to therapy, and that in the cases it takes on it is able to succeed without making mistakes, the present treatise demonstrates, as do the explanations of

---

[36] ᾗ Jouanna: ἣ A: om. M

[37] μὴ om. M

χνην ἐπιδείξιες, ἃς ἐκ τῶν ἔργων ἥδιον ἢ ἐκ τῶν λόγων[38] *ἐπιδεικνύουσιν, οὐ τὸ λέγειν καταμελετήσαντες, ἀλλὰ τὴν πίστιν τῷ πλήθει ἐξ ὧν ἂν ἴδωσιν οἰκειοτέρην ἡγεύμενοι ἢ ἐξ ὧν ἂν ἀκούσωσιν.*

[38] ἥδιον . . . λόγων om. A

those knowledgeable in the art, explanations they prefer to draw from facts rather than from words, not being proficient in argument, but believing that the trust of the public is more commonly won from what they see, than from what they hear.

BREATHS

INTRODUCTION

This speech, through its able application of the rhetorical art, achieves a level of expository clarity and linguistic grace uncommon in the Collection.[1] The author argues persuasively that breath (πνεῦμα)—inside the body wind (φῦσα), and outside it air (ἀήρ)—is the cause of all diseases, which are in fact the same disease, differing in appearance due only to their particular locations in the body. While the importance of breath in the healthy and ill body is acknowledged in many Hippocratic treatises, *Breaths'* insistence on its supreme significance is unique.

The text is structured in fifteen chapters as follows:

1–5	Introduction: overview of medicine's ways of thought and practice; just as air is the dominant force in the universe, breath in the living body is its most important nutriment; breathing is essential for life; all diseases are caused by an abnormal quantity or quality of breath.
5–8	Febrile diseases are either pestilential, caused by airs hostile to a particular species, or individual, caused by a bad regimen; how breath causes the symptoms of these conditions: e.g., chills, shivering, yawning, warmth, sweating.

[1] Cf. Jouanna *Flat.*, pp. 10–24.

9–14 The role of breath in various specific diseases: e.g., of the abdomen, head, and throat; ruptures of tissues; dropsy; paralyses; the sacred disease.

15 Conclusion.

Breaths is included in Erotian's Hippocratic census, among the "Etiological and Scientific" titles, and at least two terms from the text are handled in his *Glossary*.[2] There are also testimonies to the work in many other ancient writers, in particular to the therapeutic generalization of the first chapter: "opposites are cured by opposites: for medicine consists of subtraction and addition: subtraction of what is in excess, and addition of what is missing."[3]

The treatise is known to some Arabic medical writers and may have been translated into Arabic by Ḥunain.[4] In the fifteenth century it was translated into Latin twice:

Lat. Phil. (inc. *Sunt artes quaedam, quae possidentibus*) by Francesco Philelfo from the manuscript Scorialensis Graecus 231 (Φ III,12);

Lat. Lasc. (inc. *Quaedam artes sunt, quae laborem artificibus*) by Janus Lascaris from the manuscript I in its corrected state. This translation was printed, although rarely, in the sixteenth century.[5]

[2] See Nachmanson, p. 324f.: O12 (ὁλκάδες) and X6 (χαραδρωθῶσιν), for certain, and possibly M8 (μεμιασμένον) and A43 (ἀήρ).

[3] See *Testimonien* vol. I, pp. 251–58; vol. II,1, pp. 281–87; vol. II,2, pp. 224–30; vol. III, pp. 224–36.

[4] Cf. Ullmann, p. 32; Sezgin, p. 46.

[5] The full text of the two Latin translations is printed in Nel-

The complete Greek text of *Breaths* is transmitted in the two independent Greek manuscripts M and A, and a further independent text of most of the first chapter is contained in the manuscript Vaticanus Urbinas Graecus 64 (= Urb).[6] The papyrus Mediolanensis inv. 71.77 (III AD) (= Π_{14}) from Oxyrhynchus contains text material from chapters 8, 9, and 10 of *Breaths*, which casts light on the migration of variant readings among witnesses in the period before our oldest manuscripts.[7] The pseudo-Galenic compilation of Hippocratic texts *Causes of Affections* (= *Caus. Aff.*) edited by G. Helmreich from Marcianus Venetus Graecus App. V,12 (XI/XIV c.) includes four passages from chapters 6 to 9 of *Breaths* containing important textual variants.[8]

Breaths is contained in all the collected Hippocratic editions and translations, including Zwinger, Heiberg, and Diller *Schr.*, and is the subject of two comprehensive twentieth-century studies:

Nelson, A. *Die hippokratische Schrift* ΠΕΡΙ ΦΥΣΩΝ. Uppsala, 1909. (= Nelson)

Jouanna, J. *Hippocrate. Des Vents. . . .* Budé V (1). Paris, 1988. (= Jouanna *Flat.*)

The present edition depends mainly on Jouanna's.

---

son, pp. 5–33. On their history and importance, see Nelson, pp. 50–59; Kibre, pp. 168f.; Jouanna *Flat.*, pp. 70–73.

[6] Cf. Jouanna *Flat.*, p. 52.

[7] Cf. Marganne, p. 205; Jouanna *Flat.*, pp. 68–70; *Index Hipp.*, p. xii.

[8] Cf. *Caus. Aff.*, pp. 3–19; Jouanna *Flat.*, pp. 78–83; *Testimonien* vol. II,2, p. 229f.

ΠΕΡΙ ΦΥΣΩΝ

VI 90 Littré

1. Εἰσί τινες τῶν τεχνέων, αἳ τοῖσι μὲν κεκτημένοισίν εἰσιν ἐπίπονοι, τοῖσι δὲ χρεωμένοισιν ὀνήϊστοι, καὶ τοῖσι μὲν δημότῃσι ξυνὸν ἀγαθόν, τοῖσι δὲ μεταχειριζομένοισί σφας λυπηραί· τῶν δὲ δὴ τοιούτων ἐστὶ τεχνέων ἣν οἱ Ἕλληνες καλέουσιν ἰητρικήν. ὁ μὲν γὰρ ἰητρὸς ὁρέει τε δεινά, θιγγάνει τε ἀηδέων, ἐπ' ἀλλοτρίῃσί τε συμφορῇσιν ἰδίας καρποῦται λύπας· οἱ δὲ νοσέοντες ἀποτρέπονται[1] διὰ τὴν τέχνην τῶν μεγίστων κακῶν, νούσων, λύπης, πόνων, θανάτου· πᾶσι γὰρ τούτοισιν ἄντικρυς ἡ ἰητρική. ταύτης δὲ τῆς τέχνης τὰ μὲν φλαῦρα χαλεπὸν γνῶναι, τὰ δὲ σπουδαῖα ῥηΐδιον· καὶ τὰ μὲν φλαῦρα τοῖσιν ἰητροῖσι μούνοισιν ἔστιν εἰδέναι, καὶ οὐ τοῖσι δημότῃσιν· οὐ γὰρ σώματος, ἀλλὰ γνώμης ἐστὶν ἔργα. ὅσα μὲν γὰρ χειρουργῆσαι χρή, συνεθισθῆναι δεῖ—τὸ γὰρ ἔθος τῇσι χερσὶ κάλλιστον διδασκαλεῖον[2]—περὶ δὲ τῶν ἀφανεστάτων καὶ χαλεπωτάτων νοσημάτων δόξῃ μᾶλλον ἢ τέχνῃ κρίνεται· διαφέρει δ' ἐν αὐτοῖσι πλεῖστον ἡ πείρη τῆς ἀπειρίης. | ἓν δὲ δή τι τῶν τοιούτων ἐστὶ τόδε· τί ποτε τὸ αἴτιόν ἐστι τῶν νούσων, καὶ τίς ἀρχὴ

92

BREATHS

1. There are arts which are strenuous for their possessors, but most beneficial for their users, to the public a common good, but to their practitioners a source of grief: among such arts is the one the Greeks call medicine. For the physician sees what is terrible, touches what is disgusting, and from the calamities of others reaps sorrows all his own. But the ill, by this art, avert the greatest of evils—diseases, griefs, pains and death—for medicine acts against all of these. The less understood parts of this art are difficult to know, while the more perfected parts are easier: the less sure parts are accessible to the physician alone, but not to the layman, since they are matters not of the body, but of understanding. Cases that require manual intervention demand practice—for habit is the best teacher of the hands—but the most occult and difficult diseases are to be judged more through insight than by routine technique: in these, having experience makes a very great difference from lacking it. One important question regarding such diseases is the following. What in fact is the cause of the diseases, and what is the beginning and

[1] ἀποτρέπονται A Urb: ἀπαλλάσσονται M

[2] Add. γίνεται A: γίνεται· περὶ ἀφανῶν νοσημάτων Urb

καὶ πηγὴ γίνεται τῶν ἐν τῷ σώματι κακῶν; εἰ γάρ τις εἰδείη τὴν αἰτίην τοῦ νοσήματος, οἷός τ' ἂν εἴη τὰ συμφέροντα προσφέρειν τῷ σώματι ἐκ τῶν ἐναντίων ἐπιστάμενος τῷ νοσήματι·[3] *αὕτη γὰρ ἡ ἰητρικὴ μάλιστα κατὰ φύσιν ἐστίν. αὐτίκα γὰρ λιμὸς νοῦσός ἐστιν· ὃ γὰρ ἂν λυπῇ τὸν ἄνθρωπον, τοῦτο καλέεται νοῦσος. τί οὖν λιμοῦ φάρμακον; ὃ παύει λιμόν· τοῦτο δ' ἐστὶ βρῶσις· τούτῳ ἄρα ἐκεῖνο ἰητέον. αὖτις αὖ δίψαν ἔπαυσε πόσις· πάλιν αὖ πλησμονὴν ἰῆται κένωσις, κένωσιν δὲ πλησμονή· πόνον δὲ ἀπονίη, ἀπονίην δὲ πόνος.*[4] *ἑνὶ δὲ συντόμῳ λόγῳ, τὰ ἐναντία τῶν ἐναντίων ἐστὶν ἰήματα· ἰητρικὴ γάρ ἐστιν ἀφαίρεσις καὶ πρόσθεσις, ἀφαίρεσις μὲν τῶν πλεοναζόντων, πρόσθεσις δὲ τῶν ἐλλειπόντων. ὁ δὲ τοῦτ' ἄριστα ποιέων ἄριστος ἰητρός· ὁ δὲ τούτου πλεῖστον ἀπολειφθεὶς πλεῖστον ἀπελείφθη τῆς τέχνης.*[5] *ταῦτα μὲν οὖν ἐν παρέργῳ τοῦ λόγου τοῦ μέλλοντος εἴρηται.*

2. *Τῶν δὲ δὴ νούσων ἁπασέων ὁ μὲν τρόπος ὡὐτός, ὁ δὲ τόπος διαφέρει· δοκέει μὲν οὖν οὐδὲν ἐοικέναι τὰ νοσήματα ἀλλήλοισι διὰ τὴν ἀλλοιότητα τῶν τόπων, ἔστι δὲ μία πασέων νούσων καὶ ἰδέη καὶ αἰτίη ἡ αὐτή. ταύτην δὲ ἥτις ἐστὶ διὰ τοῦ μέλλοντος λόγου φράσαι πειρήσομαι.*

3. *Τὰ σώματα καὶ τὰ τῶν ἄλλων ζῴων καὶ τὰ τῶν ἀνθρώπων ὑπὸ τρισσῶν τροφέων τρέφεται· τῇσι δὲ* 94 *τροφῇσι τάδε | ὀνόματά ἐστιν· σῖτα, ποτά, πνεῦμα.*

[3] *ἐκ τῶν . . . νοσήματι* M Urb: om. A

source of the evils in the body? For if a person knew the cause of a disease, he could apply remedies to the body, starting out from opposite things to counteract the disease: such a medicine is most in accord with nature. For example, hunger is a disease, since anything that makes a person suffer is called a disease: what then is the remedy for hunger? Whatever puts an end to hunger, food then, and hunger should be treated with this. Or thirst, for example, is relieved by drink. Or then again fullness, which is cured by emptiness, and emptiness which is cured by fullness; just as rest relieves exhaustion, and exertion relieves idleness. In one short word, opposites are cured by opposites: for medicine consists of subtraction and addition: subtraction of what is in excess, and addition of what is missing. The person who can do this best is the best physician, whereas the one who is most lacking in this ability is most wanting in the art. These remarks, then, are made as an introduction to the discourse that will now follow.

2. The way of all diseases is consistently the same, but their location varies. Now diseases appear not to resemble one another at all, because of the difference of their sites, but in fact they all have one identical form and the same cause: what this is, I shall attempt to explain in the discourse that follows.

3. The bodies of other living beings and of humans are nourished by three nutriments; the names of these nutriments are food, drink, and breath. Breath inside bodies is

---

[4] ἀπονίῃν δὲ πόνος M: om. A Urb

[5] The text in Urb ends at this point.

πνεῦμα δὲ τὸ μὲν ἐν τοῖσι σώμασι φῦσα καλέεται, τὸ δὲ ἔξω τῶν σωμάτων ὁ ἀήρ. οὗτος δὲ μέγιστος ἐν τοῖσι πᾶσι τῶν πάντων δυνάστης ἐστίν· ἄξιον δ' αὐτοῦ θεήσασθαι τὴν δύναμιν. ἄνεμος γάρ ἐστιν ἠέρος ῥεῦμα καὶ χεῦμα· ὅταν οὖν πολὺς ἀὴρ ἰσχυρὸν ῥεῦμα ποιήσῃ, τά τε δένδρα ἀνασπαστὰ πρόρριζα γίνεται διὰ τὴν βίην τοῦ πνεύματος, τό τε πέλαγος κυμαίνεται, ὁλκάδες τε ἀπείρατοι[6] μεγέθει διαρριπτεῦνται. τοιαύτην μὲν οὖν ἐν τούτοισιν ἔχει δύναμιν· ἀλλὰ μήν ἐστί γε τῇ μὲν ὄψει ἀφανής, τῷ δὲ λογισμῷ φανερός. τί γὰρ ἄνευ τούτου γένοιτ' ἄν; ἢ τίνος οὗτος ἄπεστιν; ἢ τίνι οὐ ξυμπάρεστιν; ἅπαν γὰρ τὸ μεταξὺ γῆς τε καὶ οὐρανοῦ πνεύματος ξύμπλεόν ἐστιν. τοῦτο καὶ χειμῶνος καὶ θέρεος αἴτιον, ἐν μὲν τῷ χειμῶνι πυκνὸν καὶ ψυχρὸν γινόμενον, ἐν δὲ τῷ θέρει πρηῢ καὶ γαληνόν. ἀλλὰ μὴν ἡλίου τε καὶ σελήνης καὶ ἄστρων ὁδὸς διὰ τοῦ πνεύματός ἐστιν· τῷ γὰρ πυρὶ τὸ πνεῦμα τροφή· πῦρ δὲ ἠέρος στερηθὲν οὐκ ἂν δύναιτο ζώειν· ὥστε καὶ τὸν τοῦ ἡλίου δρόμον ἀέναον ἐόντα[7] ὁ ἀὴρ ἀέναος καὶ λεπτὸς ἐὼν παρέχεται. ἀλλὰ μὴν ὅτι γε καὶ τὸ πέλαγος μετέχει πνεύματος, φανερόν· οὐ γὰρ ἄν ποτε τὰ πλωτὰ ζῷα ζώειν ἐδύνατο, μὴ μετέχοντα πνεύματος· μετέχοι δ' ἂν πῶς ἂν ἄλλως ἀλλ' ἢ διὰ τοῦ ὕδατος κἀκ τοῦ ὕδατος[8] ἕλκοντα τὸν ἠέρα; καὶ μὴν ἥ τε γῆ τούτῳ βάθρον, οὗτός τε γῆς ὄχημα, κενεόν τε οὐδέν ἐστιν τούτου. |

96 4. Διότι μὲν οὖν ἐν τοῖσιν ὅλοισιν ὁ ἀὴρ ἔρρωται, εἴρηται· τοῖσι δ' αὖ θνητοῖσιν οὗτος αἴτιος τοῦ βίου,

called wind, outside bodies, air. Air is the greatest potency ruling in all parts of all things; this potency deserves to be considered. For wind is a flux and stream of air; thus when much air creates a powerful flow, the trees are torn out by their roots due to the force of the wind, and the sea is churned into waves, and barges of vast magnitude are cast about. So much potency, then, does wind have in these things, yet it is invisible to the sight—although perceptible to the understanding. And what could take place without it? Or from what thing is it absent? And in what is it not present? For everything between earth and heaven is filled with air. It is the cause of winter and summer, being dense and cold in winter, but mild and placid in summer. Besides, the movement of the sun and the moon and the stars is through the air, for air is a nutriment for fire—cut off from air, a fire cannot live—so that air, being eternal and of fine consistency also provides eternal movement to the sun. And furthermore, it is clear that the sea contains air, since if they did not have a supply air the sea creatures would never be able to survive; and how could they acquire air other than by drawing it through the water and from the water? In fact the earth is a base for the air, and air is a vehicle for the earth; and nothing is empty of air.

4. Now that air is powerful in the whole universe has been explained. Moreover, for human beings, air is the

---

[6] ἀπείρατοι A: ἄπειροι τῷ M
[7] δρόμον ἀέναον ἐόντα M: βίον ἀένναον A
[8] διὰ τοῦ . . . ὕδατος M: τοῦ ὕδατος A

καὶ τῶν νούσων τοῖσι νοσέουσι· τοσαύτη δὲ τυγχάνει ἡ χρείη πᾶσι τοῖσι σώμασι τοῦ πνεύματος ἐοῦσα, ὥστε τῶν μὲν ἄλλων ἁπάντων ἀποσχόμενος ὤνθρωπος καὶ σιτίων καὶ ποτῶν δύναιτ᾽ ἂν ἡμέρας καὶ δύο καὶ τρεῖς καὶ πλέονας διάγειν· εἰ δέ τις ἐπιλάβοι[9] *τὰς τοῦ πνεύματος ἐς τὸ σῶμα διεξόδους,*[10] *ἐν βραχεῖ μέρει ἡμέρης ἀπόλοιτ᾽ ἄν, ὡς μεγίστης τῆς χρείης ἐούσης τῷ σώματι τοῦ πνεύματος. ἔτι τοίνυν τὰ μὲν ἄλλα πάντα διαλείπουσιν οἱ ἄνθρωποι πρήσσοντες. ὁ γὰρ βίος μεταβολέων πλέος· τοῦτο δὲ μοῦνον ἀεὶ διατελέουσιν ἅπαντα τὰ θνητὰ ζῷα πρήσσοντα, τοτὲ μὲν ἐμπνέοντα, τοτὲ δὲ ἐκπνέοντα.*[11]

5. *Ὅτι μὲν οὖν μεγάλη κοινωνίη ἅπασι τοῖσι ζῴοισι τοῦ ἠέρος ἐστίν, εἴρηται· μετὰ τοῦτο τοίνυν εὐθέως*[12] *ῥητέον, ὅτι οὐκ ἄλλοθέν ποθεν εἰκός ἐστι γίνεσθαι τὰς ἀρρωστίας ἢ ἐντεῦθεν, ὅταν τοῦτο πλέον ἢ ἔλασσον ἢ ἀθροώτερον γένηται ἢ μεμιασμένον νοσηροῖσι μιάσμασιν ἐς τὸ σῶμα ἐσέλθῃ. περὶ μὲν οὖν ὅλου τοῦ πρήγματος ἀρκέει μοι ταῦτα· μετὰ δὲ ταῦτα πρὸς αὐτὰ τὰ ἔργα τῷ αὐτῷ λόγῳ πορευθεὶς ἐπιδείξω τὰ νοσήματα τούτου ἀπόγονά τε καὶ*[13] *ἔκγονα πάντα ἐόντα.*

6. *Πρῶτον δὲ ἀπὸ τοῦ κοινοτάτου νοσήματος ἄρξομαι, πυρετοῦ· τοῦτο γὰρ τὸ νόσημα πᾶσιν ἐφεδρεύει τοῖσιν ἄλλοισι νοσήμασιν. ἔστι δὲ δισσὰ ἔθνεα πυρετῶν, ὡς ταύτῃ διελθεῖν· ὁ μὲν κοινὸς ἅπασιν, ὁ καλεόμενος* | *λοιμός· ὁ δὲ [διὰ πονηρὴν δίαιταν]*[14] *ἰδίῃ τοῖσι πονηρῶς διαιτωμένοισι γινόμενος.*

98

cause of life, and in people who are ill it is the cause of their diseases. The need for breath is so great for all (sc. living) bodies that although a person deprived of everything else, both foods and drinks, could endure for two, three and more days, if someone had the paths of breath into his body blocked, he would die in a short part of one day: so great is the body's need of breath. Besides, human beings interrupt all their other actions—life being full of changes—but this one action all mortal living beings perform continuously, at one moment inspiring, at another moment expiring.

5. Now, that there is a great commonality of breath among all living beings has been established. Next I must immediately indicate that in all likelihood maladies have no other source from which they arise except from this, when breath is too much or too little or too gusty, or it enters the body polluted by disease miasms. On this point in general so much will suffice. Next I will move on in my account to the actual details themselves, and show that diseases are in fact all descended and produced from air.

6. I will begin first from the general disease state of fever, for this disease overlays all the other disease phenomena. To clarify, there are two species of fever, one which is shared by everyone and called a pestilence, and another which is limited to an individual, and arises from

---

9 ἐπιλάβοι M: ἀπολάβοι A
10 διεξόδ. M: ἐξόδ. A: ἐσόδ. Nelson
11 ἐμπν. . . . ἐκπν. M: ἐνπν. . . . ἀναπν. A
12 εὐθέως M: om. A
13 ἀπόγονά τε καὶ M: om. A
14 Del. Nelson

ἀμφοτέρων δὲ τούτων ὁ ἀὴρ αἴτιος. ὁ μὲν οὖν κοινὸς[15] πυρετὸς διὰ τοῦτο τοιοῦτός ἐστιν, ὅτι τὸ πνεῦμα τωὐτὸ πάντες ἕλκουσιν· ὁμοίου δὲ ὁμοίως τοῦ πνεύματος τῷ σώματι μιχθέντος, ὅμοιοι καὶ οἱ πυρετοὶ γίνονται. ἀλλ᾽ ἴσως φήσει τις· διὰ τί οὖν οὐχ ἅπασι τοῖσι ζῴοισιν, ἀλλ᾽ ἔθνει τινὶ αὐτῶν ἐπιπίπτουσιν αἱ τοιαῦται νοῦσοι; ὅτι διαφέρει, φαίην ἄν, καὶ σῶμα σώματος,[16] καὶ φύσις φύσιος, καὶ τροφὴ τροφῆς· οὐ γὰρ πᾶσι τοῖσιν ἔθνεσι τῶν ζῴων ταὐτὰ οὔτ᾽ ἀνάρμοστα οὔτ᾽ ἐνάρμοστά ἐστιν, ἀλλ᾽ ἕτερα ἑτέροισι σύμφορα, καὶ ἕτερα ἑτέροισιν ἀσύμφορα· ὅταν μὲν οὖν ὁ ἀὴρ τοιούτοισι χρωσθῇ μιάσμασιν, ἃ τῇ ἀνθρωπίνῃ φύσει πολέμιά ἐστιν, ἄνθρωποι τότε νοσέουσιν· ὅταν δὲ ἑτέρῳ τινὶ ἔθνει τῶν ζῴων ἀνάρμοστος ὁ ἀὴρ γένηται, κεῖνα τότε νοσέουσιν.

7. Αἱ μέν νυν δημόσιαι τῶν νούσων εἴρηνται, καὶ δι᾽ ὅτι καὶ ὅκως καὶ οἷσι καὶ ἀπ᾽ ὅτευ γίνονται· τὸν δὲ δὴ διὰ πονηρὴν δίαιταν γινόμενον πυρετὸν διέξειμι. πονηρὴ δέ ἐστιν ἡ τοιήδε δίαιτα· τοῦτο μὲν ὅταν τις πλέονας τροφάς, ὑγρὰς ἢ ξηράς, διδοῖ τῷ σώματι ἢ τὸ σῶμα δύναται φέρειν, καὶ πόνον μηδένα τῷ πλήθει τῶν τροφέων ἀντιτιθῇ, τοῦτο δ᾽ ὅταν ποικίλας καὶ ἀνομοίους ἀλλήλῃσιν ἐσπέμπῃ τροφάς· τὰ γὰρ ἀνόμοια στασιάζει, καὶ τὰ μὲν θᾶσσον, τὰ δὲ σχολαίτερον πέσσεται. μετὰ δὲ πολλῶν σιτίων ἀνάγκη καὶ

100 πολὺ | πνεῦμα ἐσιέναι· μετὰ πάντων γὰρ τῶν ἐσθιομένων τε καὶ πινομένων ἀπέρχεται πνεῦμα ἐς τὸ

following a bad regimen. Air is the cause of both. The fever common to all is like this because everyone inspires the same air: as the same air is mixed with each body in the same way, the same fevers too are engendered. But perhaps someone will ask: "But why then do not such diseases attack all kinds of animals, but only a certain species of them?" To this I would reply: "Because they (sc. the species of animals) differ from body to body, from nature to nature, and from nutriment to nutriment. In the different species of living beings it is not the same things that are inappropriate or appropriate, but some things are beneficial to some species, and other things are beneficial to others, and the same holds true for harmful things. Thus, when the air becomes infected with such miasms as are hostile to human nature, human beings become ill, but when the air becomes hostile to some other species, that species becomes ill."

7. The common (sc. pestilential) diseases have now been explained as to why, how, in whom, and from what cause they arise. Next I will go through the fevers that arise from bad regimens. A bad diet is one like this: as a first example, when someone gives the body more nutriment, moist or dry, than it can bear, and does not counterbalance the quantity of nutriment with any exertion; another example is when someone ingests nutriments that are very dissimilar and unlike one another, for such different foods contend among one another, some being digested more quickly, and others more slowly. Along with excessive food, excessive breath must also enter the body,

---

[15] οὖν κοινὸς M: πολύκοινος A [16] Add. καὶ ἀὴρ ἠέρος A

σῶμα ἢ πλέον ἢ ἔλασσον. φανερὸν δ' ἐστὶν τῷδε· ἐρυγαὶ γίνονται μετὰ τὰ σιτία καὶ τὰ ποτὰ τοῖσι πλείστοισιν· ἀνατρέχει γὰρ ὁ κατακλεισθεὶς ἀήρ, ὅταν ἀναρρήξῃ τὰς πομφόλυγας, ἐν ᾗσι κρύπτεται. ὅταν οὖν τὸ σῶμα πληρωθὲν τροφῆς πλησθῇ καὶ πνεύματος, ἐπὶ πλέον[17] τῶν σιτίων χρονιζομένων—χρονίζεται δὲ τὰ σιτία διὰ τὸ πλῆθος οὐ δυνάμενα διελθεῖν—ἐμφραχθείσης δὲ τῆς κάτω κοιλίης, ἐς ὅλον τὸ σῶμα διέδραμον αἱ φῦσαι· προσπεσοῦσαι δὲ πρὸς τὰ ἐναιμότατα τοῦ σώματος ἔψυξαν. τούτων δὲ τῶν τόπων ψυχθέντων, ὅπου αἱ ῥίζαι καὶ αἱ πηγαὶ τοῦ αἵματός εἰσι, διὰ παντὸς τοῦ σώματος φρίκη διῆλθεν· ἅπαντος δὲ τοῦ αἵματος ψυχθέντος, ἅπαν τὸ σῶμα φρίσσει.

8. Διὰ τοῦτο μέν νυν αἱ φρῖκαι γίνονται πρὸ τῶν πυρετῶν· ὅπως δ' ἂν ὁρμήσωσιν αἱ φῦσαι πλήθει καὶ ψυχρότητι, τοιοῦτο γίνεται τὸ ῥῖγος· ἀπὸ μὲν πλεόνων καὶ ψυχροτέρων ἰσχυρότερον, ἀπὸ δὲ ἐλασσόνων καὶ ἧσσον ψυχρῶν ἧσσον ἰσχυρόν. ἐν δὲ τῇσι φρίκῃσι καὶ οἱ τρόμοι τοῦ σώματος διὰ τόδε γίνονται. τὸ αἷμα φοβεόμενον τὴν παρεοῦσαν φρίκην συντρέχει καὶ διαΐσσει διὰ παντὸς τοῦ σώματος ἐς τὰ θερμότατα αὐτοῦ. καθαλλομένου δὲ τοῦ αἵματος ἐκ των ἀκρωτηρίων τοῦ σώματος ἐς τὰ σπλάγχνα, τρέμουσι. τὰ μὲν γὰρ τοῦ σώματος γίνεται πολύαιμα, τὰ δ' ἄναιμα· τὰ μὲν οὖν ἄναιμα διὰ τὴν ψῦξιν | οὐκ ἀτρεμέουσιν, ἀλλὰ σφάλλονται· τὸ γὰρ θερμὸν ἐξ αὐτῶν ἐκλέλοιπε· τὰ δὲ πολύαιμα διὰ τὸ πλῆθος τοῦ αἵματος τρέμουσιν·

102

since along with all the foods and drinks breath enters the body in a greater or less amount. This is shown by the fact that after taking foods and drinks most people are subject to eructations caused by the reflux of enclosed air, when it breaks out of the bubbles inside which it is hidden. Now when a body filled with nutriment is also filled with breath, as the food lingers for a longer time—it lingers because it is unable to move on through because of its excessive amount—the lower cavity becomes impacted, and the resulting winds run through the whole body; when these meet the most sanguineous parts of the body, they cool them. When cooling takes place in the parts where the roots and springs of the blood are located, a chill runs through the whole body; when all the blood is cooled, the whole body shivers.

8. This is why chills occur before the onset of fevers. The degree of chilling depends on the amount and coldness of the rushing winds: from more and colder winds there is stronger chilling, from a smaller amount of less cold winds, there is less chilling. During the chills, this is how the body's shivering too arises: the blood, being terrified by the chills taking place, runs together and rushes through the whole body to its hottest parts. Then, as blood flees from the extremities of the body to the inward parts, shivering follows. In this way, some parts of the body come to contain much blood, whereas other parts have none: the parts that lack blood due to the cold cannot stand still, but totter because they have lost their warmth; the parts with much blood tremble due to the large amount of their

[17] Add. *γίνεται* M

οὐ γὰρ δύναται πολλὸν γενόμενον ἀτρεμίζειν. χασμῶνται δὲ πρὸ τῶν πυρετῶν, ὅτι πολὺς ἀὴρ ἀθροισθείς, ἀθρόως ἄνω διεξιών, ἐξεμόχλευσε καὶ διέστησε τὸ στόμα· ταύτῃ γὰρ εὐδιέξοδός ἐστιν. ὥσπερ γὰρ ἀπὸ τῶν λεβήτων ἀτμὸς ἀνέρχεται πολύς, ἑψομένου τοῦ ὕδατος, οὕτω καὶ τοῦ σώματος θερμαινομένου διαΐσσει διὰ τοῦ στόματος ὁ ἀὴρ συνεστραμμένος καὶ βίῃ φερόμενος. τά τε ἄρθρα διαλύεται πρὸ τῶν πυρετῶν· χλιαινόμενα γὰρ τὰ νεῦρα διίσταται. ὅταν δὲ δὴ συναλισθῇ τὸ πλεῖστον τοῦ αἵματος, ἀναθερμαίνεται πάλιν ὁ ἀὴρ ὁ ψύξας τὸ αἷμα, κρατηθεὶς ὑπὸ τῆς θέρμης. διάπυρος δὲ καὶ ἀμυδρὸς γενόμενος ὅλῳ τῷ σώματι τὴν θερμασίην ἐνηργάσατο. συνεργὸν δὲ αὐτῷ τὸ αἷμά ἐστιν· τήκεται γὰρ χλιαινόμενον καὶ γίνεται ἐξ αὐτοῦ πνεῦμα. τοῦ δὲ πνεύματος προσπίπτοντος πρὸς τοὺς πόρους τοῦ σώματος, ἱδρὼς γίνεται· τὸ γὰρ πνεῦμα συνιστάμενον ὕδωρ χεῖται, καὶ διὰ τῶν πόρων διεξελθὸν ἔξω περαιοῦται τὸν αὐτὸν τρόπον, ὅνπερ ἀπὸ τῶν ἑψομένων ὑδάτων ὁ ἀτμὸς ἐπανιών, ἢν ἔχῃ στερέωμα πρὸς ὅ τι χρὴ προσπίπτειν, παχύνεται καὶ πυκνοῦται, καὶ σταγόνες ἀποπίπτουσιν ἀπὸ τῶν πωμάτων, οἷσιν ἂν ὁ ἀτμὸς προσπίπτῃ. πόνοι δὲ κεφαλῆς ἅμα τῷ πυρετῷ γίνονται διὰ τόδε· στενοχωρίη τῇσι διεξόδοισιν ἐν τῇ κεφαλῇ γίνεται <τῇσι>[18] τοῦ αἵματος· πέπληνται γὰρ αἱ φλέβες ἠέρος· πλησθεῖσαι δὲ καὶ πρησθεῖσαι τὸν πόνον ἐμποιέουσιν ἐν τῇ κεφαλῇ· βίῃ γὰρ τὸ | αἷμα βιαζόμενον διὰ στενῶν ὁδῶν θερμὸν ἐὸν οὐ δύναται περαιοῦ-

104

blood, since the blood, from its increase, cannot remain still. Yawning befalls the patient before the onset of fever, because much air collects together, and, by rushing up out of the mouth, acts like a lever and forces it open, since that is the point through which it can most easily escape. Just as when much steam forces its way out of a kettle containing boiling water, air in the heated body rushes up and out of the mouth as it is compressed and expelled by force. The joints, for their part, become looser before the onset of fevers, as their cords are warmed and slackened. But when the majority of the blood has been collected together, the air that first cooled it is overcome by its heat and warms back up: as this becomes inflamed and diffused, it brings heat to the whole body, aided by the blood which, as it warms, melts and emits breath. When this breath arrives at the body's pores, sweat is exuded: for, as the breath condenses, liquid is expelled and comes out of the pores in the same way that steam rising up from boiling water, on encountering some solid object, thickens, condenses and makes drops form hanging down from the (sc. vessels') lids where the steam strikes them. Troubles in the head accompanying fevers arise as follows: space for the passage of blood in the head becomes narrow, as the vessels there are filled up with air; this narrowing causes the vessels to become full and swollen, which provokes headache, as the blood is pressed forcefully through the narrowed passageways, and, despite its hotness, cannot

---

[18] Add. Jouanna from *Caus. Aff.* ‹ταῖς›

σθαι ταχέως· πολλὰ γὰρ ἐμποδὼν αὐτῷ κωλύματα καὶ ἐμφράγματα· διὸ δὴ καὶ οἱ σφυγμοὶ γίνονται περὶ τοὺς κροτάφους.

9. Οἱ μὲν οὖν πυρετοὶ διὰ ταῦτα γίνονται καὶ τὰ μετὰ τῶν πυρετῶν ἀλγήματα καὶ νοσήματα· τῶν δὲ ἄλλων ἀρρωστημάτων, οἱ μὲν εἰλεοὶ ἢ ἀνειλήματα ἢ στρόφοι[19] ἢ ἀποστηρίγματα ὅτι φυσέων ἐστί, πᾶσιν ἡγέομαι φανερὸν εἶναι. πάντων γὰρ τῶν τοιούτων μία ἰητρική, τοῦ πνεύματος ἀπαρύσαι. τοῦτο γὰρ ὅταν προσπέσῃ πρὸς τόπους ἀπαθέας καὶ ἁπαλοὺς καὶ ἀήθεας καὶ ἀθίκτους,[20] ὥσπερ τόξευμα ἐγκείμενον διαδύνει διὰ τῆς σαρκός· προσπίπτει δὲ τοτὲ μὲν πρὸς τὰ ὑποχόνδρια, τοτὲ δὲ πρὸς τὰς λαπάρας, τοτὲ δὲ ἐς ἀμφότερα. διὸ δὴ καὶ θερμαίνοντες ἔξωθεν πυριήμασι πειρέονται μαλθάσσειν τὸν τόπον· ἀραιούμενον γὰρ ὑπὸ τῆς θερμασίης τοῦ πυριήματος διέρχεται τὸ πνεῦμα διὰ[21] τοῦ σώματος, ὥστε παῦλάν τινα γενέσθαι τῶν πόνων.

10. Ἴσως δ' ἄν τις εἴποι· πῶς οὖν καὶ τὰ ῥεύματα γίνεται διὰ τὰς φύσας; ἢ τίνα τρόπον τῶν αἱμορραγιῶν τῶν περὶ τὰ στέρνα τοῦτ' αἴτιόν ἐστιν; οἶμαι δὲ καὶ ταῦτα δηλώσειν διὰ τοῦτο γινόμενα. ὅταν αἱ περὶ τὴν κεφαλὴν φλέβες γεμισθῶσιν ἠέρος, πρῶτον μὲν ἡ κεφαλὴ βαρύνεται, τῶν φυσέων ἐγκειμένων· ἔπειτα εἰλεῖται τὸ αἷμα, οὐ διαχέειν δυναμένων[22] διὰ τὴν στενότητα τῶν ὁδῶν· τὸ δὲ λεπτότατον τοῦ αἵματος

[19] ἢ στρόφοι M: om. A

progress rapidly due to many obstructions and barriers impeding it: for the same reason, throbbing in the temples also follows.

9. This, then, is how fevers arise, and the troubles and diseases that accompany them. Among other diseases, the fact that ileus, volvulus, colics and fixed pains arise from winds is, I believe, clear to everyone. For all these share the same treatment: *viz.* to draw off wind. When such winds are directed against regions that are sound, tender, unaccustomed, and untouched, they penetrate through the flesh like an arrow; sometimes they strike the hypochondrium, sometimes the flanks, and sometimes both of these. For this reason one attempts, by applying warmth from the outside in the form of fomentations, to soften the area; for as it is rarefied by the warmth of a fomentation, the wind is dispersed through the body, so that the pains are somewhat relieved.

10. But perhaps someone will ask: "But how can fluxes too be caused by winds?" or "In what way do winds bring about hemorrhages in the chest?" Well, I think I can show that these too have the same cause. When the vessels of the head inflate with air, first the head becomes heavy due to the winds occupying it, and then the blood there becomes turbulent because the narrowing of the passages prevents the winds from pouring the blood through: the finest part of the blood, however, is squeezed out through

---

[20] *ἀπαθέας . . . ἀθίκτους* Jouanna: *ἀπαλοὺς καὶ ἀήθεας καὶ ἀθίκτους* (with the marginal variant reading *εὐθίκτους* to *ἀθίκτους*) M: *ἀπαθέας καὶ ἀήθεας* A [21] *διὰ* M: om. A

[22] *οὐ . . . δυναμένων* A: om. M

διὰ τῶν φλεβῶν ἐκθλίβεται. τοῦτο δὴ τὸ ὑγρὸν ὅταν ἀθροισθῇ πολύ, ῥεῖ δι' ἄλλων πόρων· ὅπῃ δ' ἂν
106 ἀθρόον ἀφίκηται τοῦ σώματος, | ἐνταῦθα συνίσταται ἡ νοῦσος· ἢν μὲν οὖν ἐπὶ τὴν ὄψιν ἔλθῃ, ταύτῃ ὁ πόνος· ἢν δὲ ἐς τὰς ἀκοάς, ἐνταῦθα ἡ νοῦσος· ἢν δ' ἐς τὰς ῥῖνας, κόρυζα·[23] ἢν δὲ ἐς τὰ στέρνα, βράγχος καλέεται. τὸ γὰρ φλέγμα δριμέσι χυμοῖσι μεμιγμένον, ὅπῃ ἂν προσπέσῃ ἐς ἀήθεας τόπους, ἑλκοῖ· τῇ δὲ φάρυγγι ἁπαλῇ ἐούσῃ ῥεῦμα προσπῖπτον τρηχύτητας ἐμποιέει· τὸ γὰρ πνεῦμα τὸ διαπνεόμενον διὰ τῆς φάρυγγος ἐς τὰ στέρνα βαδίζει,[24] καὶ πάλιν ἐξέρχεται διὰ τῆς ὁδοῦ ταύτης. ὅταν οὖν ἀπαντήσῃ τῷ ῥεύματι τὸ πνεῦμα κάτωθεν ἰὸν κάτω ἰόντι, βὴξ ἐπιγίνεται, καὶ ἀναρρίπτεται[25] ἄνω τὸ φλέγμα· τούτων δὲ τοιούτων ἐόντων ἡ φάρυγξ ἑλκοῦται καὶ τρηχύνεται καὶ θερμαίνεται καὶ ἕλκει τὸ ἐκ τῆς κεφαλῆς ὑγρόν, θερμὴ ἐοῦσα· ἡ δὲ κεφαλὴ παρὰ τοῦ ἄλλου σώματος λαμβάνουσα τῇ φάρυγγι διδοῖ. ὅταν οὖν ἐθισθῇ τὸ ῥεῦμα ταύτῃ ῥέειν καὶ χαραδρωθέωσιν οἱ πόροι, διαδιδοῖ ἤδη καὶ ἐς τὰ στέρνα· δριμὺ δ' ἐὸν τὸ φλέγμα προσπῖπτόν τε τῇ σαρκὶ ἑλκοῖ, καὶ ἀναρρηγνύει τὰς φλέβας· ὅταν δὲ ἐκχυθῇ τὸ αἷμα εἰς ἀλλότριον τόπον,[26] χρονιζόμενον καὶ σηπόμενον γίνεται πύον· οὔτε γὰρ ἄνω δύναται ἀνελθεῖν οὔτε κάτω ὑπελθεῖν· ἄνω μὲν γὰρ οὐκ εὔπορος ἡ πορείη πρὸς ἄναντες ὑγρῷ χρήματι πορεύεσθαι, κάτω δὲ κωλύει ὁ φραγμὸς τῶν φρενῶν. διὰ τί δὲ δήποτε τὸ αἷμα τὸ ἄνευ ῥεύματος

the vessels. Then, when much of this moisture collects, it runs through other passages: wherever in the body this mass arrives, a disease arises. If, for example, it goes to an eye, the trouble is there; if to the ears, the disease is there; if to the nostrils, it will be a coryza; if to the chest, it is called a sore throat. For phlegm mixed with acrid humors causes ulcerations wherever it arrives in unaccustomed locations. Since the throat is tender, the invading flux causes roughness, as the breath being inhaled through the throat into the chest enters and then returns back along the same path. Now when the breath being exhaled from below encounters the flux moving down from above, a cough results, and the phlegm is expectorated upward. With these things happening, the throat ulcerates, becomes rough, and is heated, and being heated draws moisture from the head: this moisture the head takes from the rest of the body and supplies to the throat. Once this flux to the throat becomes established, and its passageways open up into wide channels, from then on the flux passes on to the chest. The phlegm, being acrid and assailing the tissues, ulcerates them, and breaks open the vessels, and when blood seeps out of them into an alien region, after a time it putrefies and turns to pus, since it cannot move either up or down: to pass upward is impossible since liquids do not easily move in that direction; in the downward direction, the diaphragm acts as an impediment. But how can the blood that breaks out ever do this when

---

[23] ἢν δ'ἐς . . . κορύζα M: om. A
[24] βαδίζει M: πορεύεται A
[25] Add. ἐς τὰ A [26] εἰς . . . τόπον M: om. A

ἀναρρηγνύμενον[27] ἀναρρήγνυται; τὸ μὲν αὐτόματον,
108 τὸ δὲ διὰ πόνους· αὐτόματον | μὲν οὖν ὅταν αὐτόματος ὁ ἀὴρ ἐλθὼν ἐς τὰς φλέβας στενοχωρίην ποιήσῃ τῇσι τοῦ αἵματος διεξόδοισι· τότε γὰρ πιεζεύμενον τὸ αἷμα πολὺ γενόμενον ἀναρρηγνύει τοὺς πόρους, ᾗ ἂν μάλιστα βρίσῃ. ὅσοι δὲ διὰ πόνων πλῆθος ἡμορράγησαν, καὶ τούτοισιν οἱ πόνοι πνεύματος ἐνέπλησαν τὰς φλέβας· ἀνάγκη γὰρ τὸν πονέοντα τόπον κατέχειν τὸ πνεῦμα. τὰ δ᾽ ἄλλα τοῖσιν εἰρημένοισιν ὅμοια γίνεται.

11. Τὰ δὲ ῥήγματα πάντα γίνεται διὰ τόδε· ὅταν ὑπὸ βίης διαστέωσιν αἱ σάρκες ἀπ᾽ ἀλλήλων, ἐς δὲ τὴν διάστασιν ὑποδράμῃ πνεῦμα, τοῦτο τὸν πόνον παρέχει.

12. Ἢν δὲ διὰ τῶν σαρκῶν αἱ φῦσαι διεξιοῦσαι τοὺς πόρους τοῦ σώματος ἀραιοὺς ποιέωσιν, ἐν[28] δὲ τῇσι φύσησιν ὑγρασίη ᾖ τις, τὴν ὁδὸν ὁ ἀὴρ ὑπηργάσατο· διαβρόχου δὲ γενομένου τοῦ σώματος, ὑπεκτήκονται μὲν αἱ σάρκες, οἰδήματα δὲ ἐς τὰς κνήμας καταβαίνει· καλέεται[29] δὲ τὸ τοιοῦτο νόσημα ὕδρωψ. μέγιστον δὲ σημεῖον ὅτι φῦσαι τοῦ νοσήματός εἰσιν αἴτιαι, τόδ᾽ ἐστίν· ἤδη τινὲς ὀλεθρίως ἔχοντες ἐκλύσθησαν καὶ ἐκενώθησαν τοῦ ὕδατος. παραυτίκα μὲν

[27] αἷμα . . . ἀναρρηγνύμενον Jouanna *Flat.* (see p. 118, n. 6): ῥεῦμα τὸ ἄνευ πνεύματος ἀναρηγνύμενον M: ῥεῦμα A

[28] ἐν A: ἕπεται M

[29] καλέεται A: καὶ λέγεται M

its rupture occurs without any flux (sc. of phlegm)? First spontaneously, and then due to strains. It may happen spontaneously when air spontaneously enters the vessels, causing a narrowing of the passages that serve for the transmission of the blood, which causes the compressed blood to increase in amount, and where it is most powerful to tear open the pores. Hemorrhaging due to excessive strains is caused too by the vessels being filled with breath, since parts being strained must necessarily retain breath; otherwise, the process is similar to what has been described.

11. All ruptures come about when tissues separate from one another as the result of violence, and breath runs into the opening, causing this injury.

12. If winds being transmitted through the tissues expand the pores of the body, and in the winds there is moisture, the air serves to prepare the way for this moisture. As the body becomes sodden, the tissues tend to melt away, and edema descends into the lower legs: such a disease is called dropsy. The following is the surest sign that it is breaths that are causing the disease: there have been patients in a terminal state of the condition who have been drained and emptied of their water:[1] now at the be-

[1] This passage, whose textual uncertainty—*ἐκλύσθησαν* (drained) or *ἐκαύθησαν* (cauterized) from a confusion of ΛΥΣ and ΑΥ—does not affect its general narrative, refers to a paracentesis procedure also depicted in *Affections* 22 (Littré, vol. 6, p. 234); but whereas the *Affections* passage reveals a more differentiated understanding of abdominal anatomy ("the 'cavity' [i.e., gastrointestinal tract] does not transmit this water into itself, but instead it forms in the region around the cavity"), the present passage uses the term "cavity" ambiguously.

οὖν τὸ ἐξελθὸν ἐκ τῆς κοιλίης ὕδωρ πολὺ φαίνεται, χρονιζόμενον δὲ ἔλασσον γίνεται. διὰ τί οὖν γίνεται; καὶ τοῦτο δῆλον· ὅτι παραυτίκα μὲν τὸ ὕδωρ ἠέρος πλῆρές ἐστιν· ὁ δὲ ἀὴρ ὄγκον παρέχεται μέγαν· ἀπιόντος δὲ τοῦ πνεύματος ὑπολείπεται τὸ ὕδωρ αὐτό· διὸ δὴ φαίνεται μὲν ἔλασσον ἐόν, ἐστὶν δὲ ἴσον. ἄλλο δὲ αὐτῶν τόδε σημεῖον· κενωθείσης γὰρ παντελῶς τῆς
110 κοιλίης, οὐδ᾽ ἐν τρισὶν ἡμέρῃσιν εἶθ᾽ ὕστερον | πάλιν πλήρεις γίνονται. τί οὖν ἐστὶ τὸ πληρῶσαν ἀλλ᾽ ἢ πνεῦμα; τί γὰρ ἂν οὕτως ἄλλο ταχέως ἐξεπλήρωσεν; οὐ γὰρ δήπου ποτόν γε τοσοῦτον ἐσῆλθεν ἐς τὸ σῶμα. καὶ μὴν οὐδὲ σάρκες ὑπάρχουσιν ἔτι αἱ τηξόμεναι· λείπεται γὰρ ὀστέα καὶ νεῦρα καὶ ῥινός, ἀφ᾽ ὧν οὐδενὸς οὐδεμίη δύναιτ᾽ ἂν αὔξησις ὕδατος εἶναι.

13. Τοῦ μὲν οὖν ὕδρωπος εἴρηται τὸ αἴτιον. αἱ δὲ ἀποπληξίαι γίνονται διὰ τὰς φύσας· ὅταν γὰρ αὗται ψυχραὶ ἐοῦσαι καὶ πολὺ[30] διαδύνουσαι ἐμφυσήσωσι τὰς σάρκας, ἀναίσθητα ταῦτα γίνεται τοῦ σώματος· ἢν μὲν οὖν ἐν ὅλῳ τῷ σώματι πολλαὶ φῦσαι διατρέχωσιν, ὅλος ὤνθρωπος ἀπόπληκτος γίνεται· ἢν δὲ ἐν μέρει τινί, τοῦτο τὸ μέρος. καὶ ἢν μὲν ἀπέλθωσιν αὗται, παύεται ἡ νοῦσος· ἢν δὲ παραμείνωσι, παραμένει. ὅτι δὲ ταῦτα οὕτως ἔχει, χασμῶνται ξυνεχῶς.[31]

14. Δοκεῖ δέ μοι καὶ τὴν ἱερὴν καλεομένην νοῦσον τοῦτ᾽ εἶναι τὸ παρεχόμενον· οἷσι δὲ λόγοισιν ἐμαυτὸν ἔπεισα, τοῖσιν αὐτοῖσι τούτοισι καὶ τοὺς ἀκούοντας πείθειν πειρήσομαι. ἡγεόμαι δὲ οὐδὲν ἔμπροσθεν οὐ-

ginning, the water coming out of the cavity appears to be abundant, but as time passes, it becomes less. Why does this happen? The answer is clear: because at the beginning the water is filled with air, and the air contributes a great deal of its volume; but as the air disperses, only the water is left behind, making it seem to decrease in bulk, although it is in fact the same. Another indication of what is happening: after the cavity is completely emptied, less than three days later patients are full again—what then except air could be filling them? And what else could fill them so fast? Certainly that much drink could not have entered the body. Nor indeed are there still tissues that are melting, since all that is left is bones, cords, and skin, none of which could produce any increased amount of water.

13. The cause of dropsy has thus been explained. Paralyses too are caused by winds: for when cold winds permeate through into the tissues in large amounts, they inflate them, making those parts of the body lose their sensation. If perchance there are many winds running through the entire body, the paralysis is of the whole body, but if the winds are confined to one part, the paralysis is only there. If these winds are dispersed, the disease ends, but if they remain, it persists. Proof that it is so is the fact that patients hold their mouth permanently open.

14. In my opinion, winds are the cause of the disease called "sacred" too. The same reasons that convinced me of this fact I will now employ to try to persuade my hearers. Now I believe that in the individual body there is no

---

[30] ψυχραὶ . . . πολὺ M: om. A
[31] ὅτι δὲ . . . ξυνεχῶς M: om. A

δενὶ εἶναι μᾶλλον τῶν ἐν τῷ σώματι ξυμβαλλόμενον[32] ἐς φρόνησιν ἢ τὸ αἷμα. τοῦτο δὲ ὅταν μὲν ἐν τῷ καθεστεῶτι μένῃ, μένει καὶ ἡ φρόνησις· ἑτεροιουμένου δὲ τοῦ αἵματος μεταπίπτει καὶ ἡ φρόνησις. ὅτι δὲ ταῦτα οὕτως ἔχει, πολλὰ τὰ μαρτυρέοντα· πρῶτον μέν, ὅπερ ἅπασι ζῴοισι κοινόν ἐστιν, ὁ ὕπνος, οὗτος μαρτυρεῖ τοῖσιν εἰρημένοισιν· ὅταν γὰρ ἐπέλθῃ τῷ σώματι, 112 τότε[33] τὸ αἷμα | ψύχεται· φύσει γὰρ ὁ ὕπνος πέφυκεν ψύχειν· ψυχθέντι δὲ τῷ αἵματι νωθρότεραι γίνονται αἱ διέξοδοι. δῆλον δέ· ῥέπει τὰ σώματα καὶ βαρύνεται—πάντα γὰρ τὰ βαρέα πέφυκεν ἐς βυσσὸν φέρεσθαι—καὶ τὰ ὄμματα συγκληΐεται, καὶ ἡ φρόνησις ἀλλοιοῦται, δόξαι τε ἕτεραί τινες ἐνδιατρίβουσιν, ἃ δὴ ἐνύπνια καλέονται. πάλιν ἐν τῇσι μέθῃσι πλέονος ἐξαίφνης γενομένου τοῦ αἵματος μεταπίπτουσιν αἱ ψυχαὶ καὶ τὰ ἐν τῇσι ψυχῇσι φρονήματα, καὶ γίνονται τῶν μὲν παρεόντων κακῶν ἐπιλήσμονες, τῶν δὲ μελλόντων ἀγαθῶν εὐέλπιδες. ἔχοιμι δ' ἂν πολλὰ τοιαῦτα εἰπεῖν, ἐν οἷσιν αἱ τοῦ αἵματος ἐξαλλαγαὶ τὴν φρόνησιν ἐξαλλάσσουσιν. ἢν μὲν οὖν παντελῶς ἅπαν ἀναταραχθῇ τὸ αἷμα, παντελῶς[34] ἡ φρόνησις ἐξαπόλλυται· τὰ γὰρ μαθήματα καὶ τὰ ἀναγνωρίσματα ἐθίσματά ἐστιν· ὅταν οὖν ἐκ τοῦ εἰωθότος ἔθεος μεταστέωμεν, ἀπόλλυται ἡμῖν ἡ φρόνησις. φημὶ δὲ τὴν ἱερὴν νοῦσον ὧδε γίνεσθαι· ὅταν πνεῦμα πολὺ κατὰ πᾶν τὸ σῶμα παντὶ τῷ αἵματι μιχθῇ, πολλὰ ἐμφράγματα γίνεται πολλαχῇ ἀνὰ[35] τὰς φλέβας. ἐπειδὰν οὖν ἐς τὰς παχείας καὶ πολυαίμους φλέβας πολὺς ἀὴρ

thing that contributes more preeminently to thought than the blood. As long as this remains in its natural state, thought is present, but if the blood becomes altered, thought too changes. Of this there are many proofs. First, something that is shared by all living beings, namely sleep, proves the truth of my argument: when sleep comes over the body, its blood is cooled, since sleep cools by nature; as this blood is cooled, its passage becomes more torpid, which is evident as the body sinks and is weighted down—for everything that is heavy tends naturally toward the depths—the eyes close, the mind wanders, and impressions of a different kind supervene, which are called dreams. In drunkenness, again, when the amount of blood suddenly increases, the mind and its thoughts change: people become forgetful of present evils and hopeful of future goods. I could also cite many other instances in which changes of the blood bring changes of thought. Thus, for example, if all the blood becomes greatly disturbed, thought is completely extinguished—for knowing and recognition are matters of habit, so that when we diverge from our normal habit, our thinking is lost. In my opinion, the sacred disease arises when much breath is mixed into all the blood through the body, creating many blockages at many points in the vessels. When much air presses into the broad vessels that contain much blood, it

---

32 -μενον Littré: -μένων MA
33 τότε M: om. A
34 ἅπαν . . . παντελῶς M: om. A
35 ἀνὰ M: κατὰ A

βρίσῃ, βρίσας δὲ μείνῃ, κωλύεται τὸ αἷμα διεξιέναι. τῇ μὲν οὖν ἐνέστηκε, τῇ δὲ νωθρῶς διεξέρχεται, τῇ δὲ θᾶσσον. ἀνομοίης δὲ τῆς πορείης τῷ αἵματι διὰ τοῦ σώματος γινομένης, παντοῖαι αἱ ἀνομοιότητες—πᾶν γὰρ τὸ σῶμα πανταχόθεν ἕλκεται καὶ τετίνακται τὰ μέρεα τοῦ σώματος ὑπηρετέοντα τῷ ταράχῳ καὶ θορύβῳ τοῦ αἵματος—διαστροφαί τε παντοῖαι παντοίως γίνονται. κατὰ δὲ τοῦτον τὸν καιρὸν ἀναίσθητοι πάντων εἰσίν, κωφοί τε τῶν λεγομένων τυφλοί τε τῶν γινομένων, ἀνάλγητοί τε πρὸς τοὺς πόνους· οὕτως ὁ ἀὴρ ταραχθεὶς ἀνετάραξε τὸ αἷμα καὶ ἐμίηνεν. ἀφροὶ 114 *δὲ διὰ τοῦ στόματος* | *ἀνατρέχουσιν εἰκότως· διὰ γὰρ τῶν σφαγιτίδων*[36] *φλεβῶν διαδύνων ὁ ἀήρ, ἀνέρχεται μὲν αὐτός, ἀνάγει δὲ μεθ' ἑωυτοῦ τὸ λεπτότατον τοῦ αἵματος· τὸ δὲ ὑγρὸν τῷ ἠέρι μιγνύμενον λευκαίνεται· διὰ λεπτῶν γὰρ ὑμένων καθαρὸς ἐὼν ὁ ἀὴρ διαφαίνεται· διὸ δὴ λευκοὶ φαίνονται παντελῶς οἱ ἀφροί. πότε οὖν παύονται τῆς νούσου καὶ τοῦ παρεόντος χειμῶνος οἱ ὑπὸ τούτου τοῦ νοσήματος ἁλισκόμενοι ἐγὼ φράσω.*[37] *ὁπόταν γυμνασθὲν ὑπὸ τῶν πόνων τὸ σῶμα θερμανθῇ, θερμαίνεται καὶ τὸ αἷμα· τὸ δὲ αἷμα διαθερμανθὲν ἐξεθέρμηνε τὰς φύσας, αὗται δὲ διαθερμανθεῖσαι διαλύονται*[38] *καὶ διαλύουσι τὴν ξύστασιν τοῦ αἵματος, αἱ μὲν συνεξελθοῦσαι μετὰ τοῦ πνεύματος, αἱ δὲ μετὰ τοῦ φλέγματος. ἀποζέσαντος δὲ τοῦ*

[36] *σφαγιτίδων* Froben: *σφαγιδίων* M: om. A
[37] *ἐγὼ φράσω* M: om. A

stays there exerting pressure, and interferes with the blood's motion: at one place the blood is held fast, at another it moves slowly forward, while at another it moves more quickly. As this irregularity of movement befalls the blood in the body, all kinds of other irregularities result—for the whole body becomes contracted everywhere, so that the individual parts are shaken, being subject to the disturbance and tumult of the blood—and there are all kinds of contortions in all kinds of ways. At that moment patients are completely unconscious, deaf to what is said, blind to what is happening, and insensible to pain from their suffering. This is how the disturbed air stirs up the blood and pollutes it. Froth also runs up through the mouth, which is natural, since air itself is coming up through the jugular vessels, and bringing with it the finest component of the blood; as this fluid is mixed with air, it turns white, since the pure air is visible through the fine membranes: for this reason the froth appears to be completely white. When it is that patients befallen by this disease will be relieved from it and the accompanying tempest, I will now explain. When the body through the effort of its exertions becomes warm, the blood too is warmed, and in becoming thoroughly heated it heats the winds. These winds, as they are thoroughly heated, themselves disperse, and they also disperse the collected blood, some winds being exhaled together with the breath, other winds in the phlegm. As the froth is boiled off, the blood

---

38 διαλύονται M: διαφέρονται A

ἀφροῦ καὶ καταστάντος τοῦ αἵματος καὶ γαλήνης ἐν τῷ σώματι γενομένης, πέπαυται τὸ νόσημα.

15. Φαίνονται τοίνυν αἱ φῦσαι διὰ πάντων τῶν νοσημάτων μάλιστα πολυπραγμονέουσαι· τὰ δ᾽ ἄλλα πάντα συναίτια καὶ μεταίτια· τὸ δὲ αἴτιον τῶν νούσων ἐὸν τοῦτο ἐπιδέδεικταί μοι. ὑπεσχόμην δὲ τῶν νούσων τὸ αἴτιον φράσειν, ἐπέδειξα δὲ τὸ πνεῦμα καὶ ἐν τοῖσιν ὅλοισι πρήγμασι δυναστεῦον καὶ ἐν τοῖσι σώμασι τῶν ζῴων· ἤγαγον δὲ τὸν λόγον ἐπὶ τὰ γνώριμα καὶ τῶν νοσημάτων καὶ[39] τῶν ἀρρωστημάτων, ἐν οἷσιν ἀληθὴς ἡ ὑπόθεσις ἐφάνη. εἰ γὰρ ἀμφὶ πάντων τῶν ἀρρωστημάτων λέγοιμι, μακρότερος μὲν ὁ λόγος ἂν γένοιτο, ἀτρεκέστερος δ᾽ οὐδαμῶς οὐδὲ πιστότερος.

[39] καὶ τῶν . . . καὶ M: om. A

quiets down, and the body returns to a peaceful state, the disease ends.

15. Thus the winds are demonstrated to be the most active agents all through the diseases; all the other determining factors are only accompanying and attendant causes. That the winds are the causes of diseases I have shown. As promised, I have explained the cause of diseases, and furthermore I have demonstrated that breath is the ruling power both of the universe as a whole, and of the bodies of living beings. I have based my argument on what is known about diseases and affections, and shown that my postulate regarding them is true. Now if I were to discuss all the ailments, my account would become much longer, but not in the least more true or more convincing.

LAW

INTRODUCTION

The subject of *Law* is medical education, discussed in a general and occasionally metaphorical way; its final chapter sets professional knowledge into a sacred context.

The text is contained in two of the primary Greek manuscripts, M and V, as well as in a third independent witness, Amb[a], which also contains the text of *Oath*. These two treatises show certain similarities of vocabulary and content.[1]

Although Erotian includes *Law* in his census of Hippocratic works, in the category "discussion of medicine," Galen a century later gives no evidence of knowing it. Through later antiquity and the middle ages the work has left only a faint trace,[2] but all the printed collected editions and translations of the Collection contain it, including Zwinger, Mack, Adams, Daremberg, Heiberg, Chadwick (under the title *Canon*), and Diller *Schr.* The present edition is dependent in particular on:

Jouanna, J. *Hippocrate. . . . La Loi*. Budé I (2). Paris, 2018. (= Jouanna *Lex*)

[1] Cf. Jouanna *Lex*, pp. 165–69.

[2] See *Testimonien* vol. I, p. 291f., and vol. III, p. 271f.; Ullmann, p. 33; Sezgin, p. 38f.; Kibre, pp. 182–86.

ΝΟΜΟΣ

IV 638 Littré

1. Ἰητρικὴ τεχνέων μὲν πασέων ἐστὶν ἐπιφανεστάτη, διὰ δὲ ἀμαθίην τῶν τε χρεομένων αὐτῇ καὶ τῶν εἰκῇ τοὺς τοιούσδε κρινόντων, πολύ τι πασέων ἤδη τῶν τεχνέων ἀπολείπεται. ἡ δὲ τῶνδε ἁμαρτὰς μάλιστά μοι δοκέει ἔχειν αἰτίην τοιήνδε· τῆς γὰρ ἰητρικῆς μούνης ἐν τῇσι πόλεσιν οὐδὲν ὥρισται πρόστιμον[1] πλὴν ἀδοξίης· αὕτη δὲ οὐ τιτρώσκει τοὺς ἐξ αὐτῆς συγκειμένους. ὁμοιότατοι γάρ εἰσιν οἱ τοιοίδε τοῖσι παρεισαγομένοισι προσώποισιν ἐν τῇσι τραγῳδίησιν· ὡς γὰρ ἐκεῖνοι σχῆμα μὲν καὶ στολὴν καὶ πρόσωπον ὑποκριτοῦ ἔχουσιν, οὐκ εἰσὶν δὲ ὑποκριταί, οὕτω καὶ οἱ ἰητροί, φήμῃ μὲν πολλοί, ἔργῳ δὲ πάγχυ βαιοί.

2. Χρὴ γάρ, ὅστις μέλλει ἰητρικὴν ξύνεσιν ἀτρεκέως ἁρμόζεσθαι, τῶνδέ μιν ἐπήβολον γενέσθαι· φύσιος, παιδομαθίης, διδασκαλίης, τόπου,[2] φιλοπονίης χρόνου. πρῶτον μὲν οὖν πάντων δεῖ φύσιος· φύσιος γὰρ

640

ἀντιπρησσούσης κενεὰ πάντα· φύσιος δὲ | ἐς τὸ ἄριστον ὁδηγεούσης, διδασκαλίῃ ἡ τέχνη[3] γίνεται, ἣν

[1] τῆς γὰρ . . . πρόστιμον Amb[a]: πρόστιμον γὰρ . . . ὥρισται MV. See Jouanna *Lex*, p. 246, n. 4.

LAW

1. Medicine is the most illustrious of all the arts, but through the lack of competence of both its practitioners and those who carelessly judge them, it is now left far behind all the other arts. This misunderstanding seems to me to be due mainly to the fact that for medicine alone no penalty is laid down in the cities, except that of ill-repute, and this cannot wound men who have no good repute to lose. Such men are like the supernumeraries in tragedies, who have the form, costume, and mask of an actor, but are not actors: like these, many men are physicians in name, but very few are physicians in deed.

2. Anyone who intends to acquire an exact medical understanding must benefit from natural inclination, education beginning from childhood, instruction, a place, industry, and time. First of all he must have the requisite nature, for if nature resists, everything will be in vain, but if nature guides the way toward what is best, the art will be established by instruction, if this is given with intelli-

---

2 *παιδομαθίης . . . τόπου* Jouanna: *παιδομαθοῦς . . . τόπου* Amb[a]: *διδασκαλίης, τρόπου εὐφυέος, παιδομαθίης* M: *παιδομαθείης, διδασκαλίης, τόπου εὐφυέος* V

3 *διδασκαλίῃ ἡ τέχνη* Amb[a]: *διδασκαλίῃ τέχνης* M: *διδασκαλίῃ τέχνη* V

μετὰ φρονήσιος ᾖ. περιποιήσασθαι δὲ παιδομαθέα γενόμενον ἐν τόπῳ ὁκοῖος εὐφυὴς πρὸς μάθησιν ἔσται· ἔτι δὲ φιλοπονίην προσενέγκασθαι ἐς χρόνον πολύν, ὅκως ἡ μάθησις ἐμφυσιωθεῖσα δεξιῶς τε καὶ εὐαλδέως τοὺς καρποὺς ἐξενέγκηται.

3. Ὁκοίη γὰρ τῶν ἐν γῇ φυομένων ἡ θεραπείη,[4] τοιήδε καὶ τῆς ἰητρικῆς ἡ μάθησις. ἡ μὲν γὰρ φύσις ἡμέων ὁκοῖον ἡ χώρη· τὰ δὲ δόγματα τῶν διδασκόντων ὁκοῖον τὰ σπέρματα· ἡ δὲ παιδομαθίη, τὸ καθ' ὥρην αὐτὰ πεσεῖν ἐς τὴν ἄρουραν· ὁ δὲ τόπος ἐν ᾧ ἡ μάθησις, ὁκοῖον ἡ ἐκ τοῦ περιέχοντος ἠέρος τροφὴ γινομένη τοῖσι φυομένοισιν· ἡ δὲ φιλοπονίη, ἐργασίη· ὁ δὲ χρόνος τολμᾷ τε ἐνισχῦσαι πάντα[5] καὶ τραφῆναι τελέως.

4. Ταῦτα ὦν χρὴ . . .[6] ἐς τὴν ἰητρικὴν τέχνην ἐσενεγκαμένους, καὶ ἀτρεκέως αὐτῆς γνῶσιν λαβόντας, οὕτως <ὡς>[7] ἀνὰ τὰς πόλιας φοιτεῦντας, μὴ λόγῳ μοῦνον, ἀλλὰ καὶ ἔργῳ ἰητροὺς νομίζεσθαι. <ἡ μὲν γὰρ>[8] ἐμπειρίη καλὸς θησαυρὸς καὶ καλὸν κειμήλιον τοῖσιν ἔχουσιν αὐτὴν καὶ ὄναρ καὶ ὕπαρ εὐθυμίης τε καὶ εὐφροσύνης πλήρης.[9] ἡ δὲ ἀπειρίη, κακὸς θησαυρὸς καὶ κακὸν κειμήλιον τοῖσιν ἔχουσιν αὐτὴν καὶ ὄναρ καὶ ὕπαρ εὐθυμίης τε καὶ εὐφροσύνης ἄμοιρος, δειλίης τε καὶ θρασύτητος τιθήνη. δειλίη μὲν γὰρ

[4] θεραπείη Potter: θεωρίη codd. Cf. Jouanna *Lex*, p. 248, n. 10.

gence. Arrange in addition that education beginning from childhood will be in a place suitable for instruction. Besides, industry must be applied for a long time, in order that the learning being instilled correctly and effectively will bear fruit.

3. The learning of medicine is like the cultivation of plants in the earth. Our natural inclination is like the soil, the ideas of our teachers are like the seeds; education beginning from childhood is like the seeds falling into the field at the right season; the location in which learning takes place is like nourishment from surrounding air being available for the plants; industry is the working of the earth; time endures for everything to become strong and reach its perfect form.

4. These are the things necessary . . . those being introduced to the medical art and acquiring an exact knowledge of it, practicing through the towns such that they are regarded as physicians not only in name, but in reality. Competence is both a fine treasure and a fine possession for those who possess it, both day and night, being full of confidence and happiness, whereas incompetence is an evil treasure and a bad possession for those who have it, both day and night, being bereft of confidence and happiness, and the nurse of fear and rashness: fear reveals a lack

---

[5] τολμᾷ τε . . . πάντα Jouanna: τολμᾷ τε ἐνισχῦσαι Amb[a]: ταῦτα ἐνισχύσει (-υσε) πάντα MV

[6] It has long been recognized that the text here is defective; Jouanna indicates a lacuna. [7] Add. Jouanna

[8] Add. Jouanna [9] ἐμπειρίη . . . πληρής Jouanna: transp. Amb[a] after τιθήνη: om. MV

642 ἀδυναμίην σημαίνει· | θρασύτης δὲ ἀτεχνίην. δύο γάρ, ἐπιστήμη τε καὶ δόξα, ὧν τὸ μὲν ἐπίστασθαι ποιέει, τὸ δὲ ἀγνοεῖν· ἡ μὲν οὖν ἐπιστήμη ποιέει τὸ ἐπίστασθαι,[10] ἡ δὲ δόξα τὸ ἀγνοεῖν.

5. Τὰ δὲ ἱερὰ ἐόντα πρήγματα ἱεροῖσιν ἀνθρώποισι δείκνυται· βεβήλοισι δὲ οὐ θέμις, πρὶν ἢ τελεσθῶσιν ὀργίοισιν ἐπιστήμης.

[10] ἡ μὲν οὖν . . . ἐπίστασθαι M Amb[a]: om. V

of ability, and rashness a lack of art. There are two kinds of knowledge: science and opinion, the former produces understanding, the latter ignorance; it is science that gives understanding, whereas opinion results in ignorance.

5. Sacred things are revealed to sacred people; to the uninitiated they are forbidden, until they have been initiated into the mysteries of science.

DECORUM

INTRODUCTION

Decorum falls into two main sections: chapters 1 to 7, an account of the medical art as a form of wisdom, its proper cultivation as opposed to its misuse, and its relationship to nature and to the gods; chapters 8 to 17, a discussion of specific medical behaviors and practices that assure the practitioner's professional status and clinical success. The final chapter, 18, summarizes the work's main message.

This text is extant in the unique independent manuscript M, although its title is included as number 58 in the index at the head of the manuscript V. No medieval reference to the work appears to exist, and the first Latin translation is that of Calvus. The treatise's text shares many features with that of *Precepts*, in particular its general obscurity and occasional incoherence, and also a common interest in medical practice and ethics. These similarities led scholars in the early twentieth century to suggest a possible identity of authorship or school affiliation for the two works,[1] but no general support for this possible relationship has emerged.

After Jones' introduction and notes to the 1923 Loeb

[1] Cf. J. F. Bensel, "Hippocratis qui fertur De medico libellus ad codicum fidem restitutus," *Philologus* 78 (1922): 88–130; Jones, vol. 2, pp. 272–76; Fleischer, pp. 105–12.

Hippocrates edition, the only important textual study devoted to the work is:

> Fleischer, U. *Untersuchungen zu den pseudohippokratischen Schriften* Παραγγελίαι, Περὶ ἰητροῦ, *und* Περὶ εὐσχημοσύνης. Berlin, 1939. (= Fleischer)

These two scholars—apparently independently, since Fleischer makes no reference to Jones—each present detailed studies of the work's style and vocabulary, identifying numerous *hapax legomena* and rare expressions, many of which occur for the first time in the Hellenistic and Roman imperial periods. These studies led both to set *Decorum's* time of composition in the second century AD and to adopt the same hypothesis as explanation for the work's textual state:

> I would insist that we must not treat *Decorum* as though it were literature. It is corrupt, but if we could restore the exact words of the writer they would still be in great part a series of ungrammatical notes to remind the lecturer of the heads of his discourse. (Jones, vol. 2, p. 271, n. 2)

> The text [of *Decorum*] is a lecture on the medical art, but from the many inconsistencies in its presentation that I have identified in my analysis, one might with more accuracy speak of the rough notes [*Konzept*] for a lecture. (Fleischer, translated from p. 107)

If, then, *Decorum*, is the record of a medical speaker's rough notes, the limits within which an editor and translator work are narrow. Neither an attempt to portray in de-

tail the innumerable difficulties of vocabulary, syntax, and interpretation the text presents, nor a literal (but largely incoherent) translation would render the general reader any great service. Thus, I have followed closely in Jones' path, accepting that in the intervening hundred years our situation has not materially changed; in fact, Jones' edition, which contains valuable notes to the text and the translation, can be consulted with profit today, and his general assessment of the situation applies equally to the present volume:

> I do not pretend, however, that the text I have printed represents the autograph, nor that the English is in many places anything but a rough paraphrase. (p. 277)

Decorum is present in the standard collected editions and translations, including Zwinger and Heiberg, as well as in a number of special studies listed by Littré (vol. 9, p. 224f.). The present edition is based on the work of earlier scholars and a collation of the manuscript M from microfilm.

ΠΕΡΙ ΕΥΣΧΗΜΟΣΥΝΗΣ

IX 226 Littré

1. Οὐκ ἀλόγως οἱ προβαλλόμενοι τὴν σοφίην πρὸς πολλὰ εἶναι χρησίμην, ταύτην δὲ τὴν ἐν τῷ βίῳ. αἱ γὰρ πολλαὶ πρὸς περιεργίην φαίνονται γεγενημέναι· λέγω δέ, αὗται αἱ μηδὲν <ἐς>[1] χρέος τῶν πρὸς ἃ διαλέγονται. ληφθείη δ' ἂν τούτων μέρεα ἐς ἐκεῖνα, ὅτι[2] οὐκ ἀργίη, οὐδὲ μὴν κακίη· τὸ γὰρ σχολάζον καὶ ἄπρηκτον ζητέει ἐς κακίην, καὶ ἀφέλκεται· τὸ δ' ἐγρηγορὸς καὶ πρός τι τὴν διάνοιαν ἐντετακὸς ἐφειλκύσατό τι τῶν πρὸς καλλονὴν βίου τεινόντων. ἐῶ δὲ[3] τούτων τὰς μηδὲν ἐς χρέος πιπτούσας διαλέξιας· χαριεστέρη γὰρ ἡ[4] πρὸς ἕτερον μέν τι ἐς τέχνην πεποιημένη, τέχνην δὲ τὴν πρὸς εὐσχημοσύνην καὶ δόξαν.

2. Πᾶσαι γὰρ αἱ μὴ μετ' αἰσχροκερδείης καὶ ἀσχημοσύνης καλαί, ᾗσι[5] μέθοδός τις ἐοῦσα τεχνικὴ ἐργάζεται· ἀλλ' εἴ γε μή, πρὸς ἀναιδείην δημεύονται.[6] νέοι τε γὰρ αὐτοῖσιν ἐμπίπτουσιν· 228 | ἀκμάζοντες δὲ δι'

[1] Littré adds ἐς on the basis of the parallel text four lines later: Fleischer understands αἱ μηδὲν χρέος (sc. ἔχουσα).

[2] ὅτι R: ἢ ὅτι M

[3] ἐῶ δὲ Littré: ἑωυτου M

[4] ἡ Littré: καὶ M

[5] κ. ᾗσι Littré: κἀκείνοισι M

DECORUM

1. It is not unreasonable for people to present wisdom[1] as a thing useful in many matters, in particular wisdom in life, although there are many wisdoms that seem to have been created frivolously, having no useful application in the areas they deal with. Some parts of this latter kind of wisdom might also be deemed acceptable, as long as they avoid idleness or indeed vice; for while leisure and idleness lead and draw people toward evil, attention and concentration of the mind on something bring with them a tendency toward the honorable life. Here I am leaving aside cases where mere talking leads to no useful result, for more pleasing is wisdom with some purpose that has been developed into an art, an art that is decorous and reputable.

2. All forms of wisdom that avoid a shameful pursuit of profit and other unseemliness are noble if they possess some artful method; otherwise, their popularity is based on impudence. Such take in young people, who then as they become mature break out in a sweat from embarrass-

[1] See *σοφία* in the Note on Technical Terms in the General Introduction to this volume.

---

[6] ἀ. δ. Ermerins: ἀναιτίην δημευταί M

ἐντροπίην ἱδρῶτας τίθενται βλέποντες· πρεσβῦται δὲ διὰ πικρίην νομοθεσίην τίθενται ἀναίρεσιν ἐκ τῶν πόλεων. καὶ γὰρ ἀγορὴν ἐργαζόμενοι οὗτοι, μετὰ βαναυσίης ἀπατέοντες, καὶ ἐν πόλεσιν ἀνακυκλέοντες οἱ αὐτοί. ἴδοι δέ τις καὶ ἐπ᾽ ἐσθῆτος καὶ ἐν τῇσιν ἄλλῃσι περιγραφῇσι· κῆν γὰρ ἔωσιν ὑπερηφανέως κεκοσμημένοι, πολὺ μᾶλλον φευκτέον καὶ μισητέον τοῖσι θεωμένοισίν ἐστιν.

3. *Τὴν δὲ ἐναντίην χρὴ ὧδε σκοπέειν· οἷς οὐ διδακτὴ κατασκευή, οὐδὲ περιεργίη· ἔκ τε γὰρ περιβολῆς καὶ τῆς ἐν ταύτῃ εὐσχημοσύνης καὶ ἀφελείης, οὐ πρὸς περιεργίην πεφυκυίης, ἀλλὰ μᾶλλον πρὸς εὐδοξίην, τό τε σύννουν, καὶ τὸ ἐν νῷ πρὸς ἑωυτοὺς διακεῖσθαι, πρός τε τὴν πορείην. οἵ τε ἑκάστῳ σχήματι τοιοῦτοι· ἀδιάχυτοι, ἀπερίεργοι, πικροὶ πρὸς τὰς συναντήσιας, εὔθετοι πρὸς τὰς ἀποκρίσιας, χαλεποὶ πρὸς τὰς ἀντιπτώσιας, πρὸς τὰς ὁμοιότητας εὔστοχοι καὶ ὁμιλητικοί, εὔκρητοι πρὸς ἅπαντας, πρὸς τὰς ἀναστάσιας σιγητικοί, πρὸς τὰς ἀποσιγήσιας ἐνθυμηματικοὶ καὶ καρτερικοί, πρὸς τὸν καιρὸν εὔθετοι καὶ λημματικοί· πρὸς τὰς τροφὰς εὔχρηστοι καὶ αὐτάρκεες, ὑπομενητικοὶ πρὸς καιροῦ τὴν*[7] *ὑπομονήν, πρὸς λόγους ἀνυστοὺς πᾶν τὸ ὑποδειχθὲν ἐκφέροντες, εὐεπίῃ χρώμενοι, χάριτι διατιθέμενοι, δόξῃ τῇ ἐκ τούτων διισχυριζόμενοι, ἐς ἀληθείην πρὸς τὸ ὑποδειχθὲν ἀποτερματιζόμενοι.*[8] |

230 4. *Ἡγεμονικώτατον μὲν οὖν τούτων ἁπάντων τῶν προειρημένων ἡ φύσις. καὶ γὰρ οἱ ἐν τέχνῃσιν, ἣν*

ment when they witness them; and as elders, in their bitterness, they pass laws banishing them out of the towns. For it is in the agora that these (sc. charlatans) operate, carrying out their deceits with vulgar tricks, circulating from town to town. You can see who they are by their clothes and their general appearance, and even if they are magnificently decked out, all the more should they be avoided and rejected by anyone who sees them.

3. The opposite (sc. useful) kind of wisdom you must look for like this. In such (sc. people) there is no studied preparation or flashiness; their dress, which is decorous and simple, aims not at show, but rather at decency, as do their thoughtfulness, concentration, and comportment. In every regard they are similar: concentrated, unruffled, precise in their argumentation, able in their replies, tenacious in opposition, clever and charming with like-minded people, good-tempered with everyone, silent in the face of disturbances, but in the face of silence persistent and patient; if given an opportunity, they know how to recognize and promptly take it; regarding nutriments they are knowledgeable about their use, and temperate; they are patient in waiting for the right moment, accomplished in expressing whatever has been revealed, good at using language, of gracious disposition, confident due to the reputation they have earned through these characteristics, and searchers after the truth in what has been revealed.

4. The guiding principle then, in each of the areas mentioned, is nature. For if those engaged in the arts act in

---

[7] *καιροῦ τὴν* Littré: *καιρόν· πρὸς* M
[8] *ἀποτερματιζόμενοι* Coray (Littré): *ἀποτελματισθῆναι* M

προσῇ αὐτοῖσι τοῦτο, διὰ πάντων τούτων πεπόρευνται τῶν προειρημένων. ἀδίδακτον γὰρ τὸ χρέος ἔν τε σοφίῃ καὶ ἐν τῇ τέχνῃ· πρόσθε μὲν ἢ διδαχθῇ, ἐς τὸ ἀρχὴν λαβεῖν ἡ δὲ φύσις κατερρύη καὶ κέχυται, τῇ δὲ σοφίῃ ἔστιν[9] *εἰδῆσαι τὰ ἀπ᾽ αὐτῆς τῆς φύσιος ποιεύμενα. καὶ γὰρ ἐν ἀμφοτέροισι τοῖσι λόγοισι πολλοὶ κρατηθέντες οὐδαμῇ συναμφοτέροισιν ἐχρήσαντο τοῖσι πρήγμασιν ἐς δεῖξιν. ἐπὴν οὖν τις αὐτῶν ἐξετάζῃ τι πρὸς ἀληθείην τῶν ἐν ῥήσει τιθεμένων, οὐδαμῇ τὰ πρὸς φύσιν αὐτοῖσι χωρήσει. εὑρίσκονται γοῦν οὗτοι παραπλησίην οἶμον ἐκείνοισι πεπορευμένοι. διόπερ ἀπογυμνούμενοι τὴν πᾶσαν ἀμφιέννυνται κακίην καὶ ἀτιμίην. καλὸν γὰρ ἐκ τοῦ διδαχθέντος ἔργου λόγος· πᾶν γὰρ τὸ ποιηθὲν τεχνικῶς ἐκ λόγου ἀνηνέχθη. τὸ δὲ ῥηθὲν τεχνικῶς, μὴ ποιηθὲν δέ, μεθόδου ἀτέχνου δεικτικὸν ἐγενήθη· τὸ γὰρ οἴεσθαι μέν, μὴ πρήσσειν δέ, ἀμαθίης καὶ ἀτεχνίης σημεῖόν ἐστιν· οἴησις γὰρ καὶ μάλιστα ἐν ἰητρικῇ αἰτίην μὲν τοῖσι κεκτημένοισιν,* 232 | *ὄλεθρον δὲ τοῖσι χρεωμένοισιν ἐπιφέρει. καὶ γὰρ ἢν ἑωυτοὺς ἐν λόγοισι πείσαντες οἰηθῶσιν εἰδέναι ἔργον τὸ ἐκ μαθήσιος, καθάπερ χρυσὸς φαῦλος ἐν πυρὶ κριθεὶς τοιούτους αὐτοὺς ἀπέδειξεν· καίτοι γε τοιαύτη ἡ πρόρρησις ἀπαρηγόρητον. ᾗ ξύνεσις ὁμογενής ἐστιν,*[10] *εὐθὺ τὸ πέρας ἐμήνυσε γνῶσις. τῶν δ᾽ ὁ χρόνος τὴν τέχνην ἐς εὐοδίην*[11] *κατέστη-*

[9] *ἔστιν* Heidel (p. 198): *ἐς τὸ* M

[10] *ᾗ . . . ὁμογενής ἐστιν* Littré: *ἐς . . . ὁμογενέσιν ὡς* M

[11] *ἐς εὐοδίην* Jones in footnote: *εὐαδέα* M

harmony with nature, they make progress in all the ways explained. For use is not a thing that can be taught in either wisdom or art, since before it is taught nature has already made the beginning by rushing down in a flood, and it is up to wisdom to learn what is done by nature itself. Many people who have come off badly in these two realms (sc. of nature and wisdom), do not apply either of them at all for demonstration in actual things. Thus when such a person examines some point being discussed with regard to truth, he will be completely at odds with its nature. These (sc. misguided practitioners) are discovered, in any case, to be following the same path as the others (i.e., the charlatans described in chapter 2 above). Thus, when they are stripped (of competence), they clothe themselves in every evil and dishonor. Reason that comes from the teaching of practical experience is a good thing, since everything that is done in accord with the art has its origin in reason. But when a thing is described as belonging to an art, but not practiced, it is clearly coming from a method devoid of art: to think, but not do, is a sign of lack of learning and art. Thought (sc. in isolation from action), especially in medicine, brings blame to practitioners, and is fatal to those who employ it. For if on the basis of mere thinking practitioners persuade themselves that they understand an art that depends on practical learning, they expose themselves, just as false gold is revealed by fire: in fact, such an exposure is inevitable. But where understanding is born together with action, knowledge immediately reveals what is to be done. In some cases, time has also set art on the correct path, and made the beginnings

σεν, ἣ τοῖσιν ἐς τὴν παραπλησίην οἶμον ἐμπίπτουσι τὰς ἀφορμὰς δήλους ἐποίησε.

5. Διὸ δὴ ἀναλαμβάνοντα τούτων τῶν προειρημένων ἕκαστα, μετάγειν τὴν σοφίην ἐς τὴν ἰητρικὴν καὶ τὴν ἰητρικὴν ἐς τὴν σοφίην. ἰητρὸς γὰρ φιλόσοφος ἰσόθεος· <οὐ>[12] πολλὴ γὰρ διαφορὴ ἐπὶ τὰ ἕτερα. καὶ γὰρ ἔνι τὰ πρὸς σοφίην ἐν ἰητρικῇ πάντα, ἀφιλαργυρίη, ἐντροπή, ἐρυθρίησις, καταστολή, δόξα, κρίσις, ἡσυχίη, ἀπάντησις, καθαριότης, γνωμολογίη, εἴδησις τῶν πρὸς βίον χρηστῶν καὶ ἀναγκαίων, |
234 † καθαρσίης ἀπεμπόλησις, † ἀδεισιδαιμονίη, ὑπεροχὴ θείη· ἔχουσι γὰρ ἃ ἔχουσι πρὸς ἀκολασίην, πρὸς βαναυσίην, πρὸς ἀπληστίην, πρὸς ἐπιθυμίην, πρὸς ἀφαίρεσιν, πρὸς ἀναιδείην. αὕτη γὰρ γνῶσις τῶν προσιόντων καὶ χρῆσις τῶν πρὸς φιλίην, καὶ ὡς καὶ ὁκοίως τὰ πρὸς τέκνα, πρὸς χρήματα. ταύτῃ μὲν οὖν ἐπικοινωνὸς σοφίη τις, ὅτι καὶ ταῦτα καὶ τὰ πλεῖστα ὁ ἰητρὸς ἔχει.

6. Καὶ γὰρ μάλιστα ἡ περὶ θεῶν εἴδησις ἐν νόῳ αὐτῇ ἐμπλέκεται. ἐν γὰρ τοῖσιν ἄλλοισι πάθεσι καὶ ἐν συμπτώμασιν εὑρίσκεται τὰ πολλὰ πρὸς θεῶν ἐντίμως κειμένη ἡ ἰητρική· οἱ δὲ ἰητροὶ θεοῖσι παρακεχωρήκασιν· οὐ γὰρ ἔνι περιττὸν ἐν αὐτῇ τὸ δυναστεῦον· καὶ γὰρ οὗτοί πολλὰ μὲν μεταχειρέονται, πολλὰ δὲ καὶ κεκράτηται αὐτοῖσι δι' ἑωυτῶν. † ἃ δὲ καταπλεονεκτεῖ νῦν ἡ ἰητρική, ἐντεῦθεν παρέξει· τὶς γὰρ ὁδὸς τῆς ἐν σοφίῃ ὧδε, καὶ γὰρ αὐτοῖσιν ἐκείνοι-

clear to those who had discovered a similar route by chance.

5. Therefore, taking up again each of the points already discussed, carry over wisdom into medicine, and medicine into wisdom—for the physician who is a lover of wisdom is the equal of a god—since there is no great difference between the two. All the features of wisdom are in fact also present in medicine: indifference to money, modesty, a sense of shame, restraint, sound opinion, judgment, reserve, firmness, purity, elevated speech, knowledge of what is useful and necessary in life, the dispensation of cleansing, freedom from superstition, divine elevation: these have what they possess in opposition to licentiousness, vulgarity, greed, passion, robbery, and shamelessness. This is knowing one's resources, the way of cultivating friendship, and the manner and means of managing one's children and one's money. Now with medicine wisdom is a kind of partner, since the physician possesses most of the same things.

6. In medicine it is especially a knowledge of the gods that is interwoven with the mind. For in other events, and especially in accidental ones, medicine is found to benefit from the honor of the gods, physicians being delivered over to the realm of the gods, since the power in medicine itself is not that great; for although they (i.e., physicians) play a healing role in many cases, many other (sc. diseases) are overcome spontaneously for them. What medicine has now mastered, from that it will draw, for there is some method of medicine in wisdom in this way, and in the

[12] Add. Zwinger in text, following Cornarius' *neque*.

σιν. † οὕτω δ' οὐκ οἴονται, ὁμολογέουσιν ὧδε τὰ περὶ σώματα παραγινόμενα· ἃ δὴ διὰ πάσης αὐτῆς πεπόρευται, μετασχηματιζόμενα ἢ μεταποιούμενα, ἃ δὲ μετὰ χειρουργίης ἰώμενα, ἃ δὲ βοηθούμενα, θεραπευόμενα | ἢ διαιτώμενα. τὸ δὲ κεφαλαιωδέστατον ἔστω ἐς τὴν τούτων εἴδησιν.

236

7. Ὄντων οὖν τοιούτων τῶν προειρημένων ἁπάντων, χρὴ τὸν ἰητρὸν ἔχειν τινὰ εὐτραπελίην παρακειμένην· τὸ γὰρ αὐστηρὸν δυσπρόσιτον καὶ τοῖσιν ὑγιαίνουσι καὶ τοῖσι νοσέουσιν. τηρεῖν δὲ χρὴ ἑωυτὸν ὅτι μάλιστα, μὴ πολλὰ φαίνοντα τῶν τοῦ σώματος μερέων, μηδὲ πολλὰ λεσχηνευόμενον τοῖσιν ἰδιώτῃσιν, ἀλλὰ τἀναγκαῖα· νομίζει γὰρ τοῦτο βίῃ εἶναι ἐς πρόκλησιν[13] θεραπηίης· ποιέειν δὲ κάρτα μηδὲν περιέργως αὐτῶν, μηδὲ μετὰ φαντασίης. ἐσκέφθω δὲ ταῦτα πάντα, ὅκως ᾖ σοι προκατηρτισμένα ἐς τὴν εὐπορίην, ὡς δέοι· εἰ δὲ μή, ἐπὶ τοῦ χρέους ἀπορίη αἰεὶ δεῖ.

8. Μελετᾶν δὲ χρὴ ἐν ἰητρικῇ ταῦτα μετὰ πάσης καταστολῆς, περὶ ψηλαφίης, καὶ ἐγχρίσιος, καὶ ἐγκαταντλήσιος, πρὸς τὴν εὐρυθμίην τῶν χειρῶν, περὶ τιλμάτων, περὶ σπληνῶν, περὶ ἐπιδέσμων, περὶ τῶν ἐκ καταστάσιος. περὶ φαρμάκων, ἐς τραύματα καὶ ὀφθαλμικά—καὶ τούτων τὰ πρὸς τὰ γένεα—ἵν' ᾖ σοι προκατηρτισμένα ὄργανά τε καὶ μηχαναὶ καὶ σίδηρος καὶ τὰ ἑξῆς· ἡ γὰρ ἐν τούτοισιν ἀπορίη ἀμηχανίη καὶ βλάβη. ἔστω δέ σοι ἑτέρη παρέξοδος ἡ λιτοτέρη

[13] βίῃ . . . πρόκλησιν Littré: πρόσκλησιν . . . βίη M

things themselves. People do not think like this, but what happens in bodies confirm it, namely what takes place in every part of medicine, due to the changes of form and quality, and what is cured by surgery, or helped by medication or diet. Let this be my very succinct summary of the knowledge particular to these things.

7. All these matters being as I have described them, it is clear that a physician must possess a certain ready wit, since dourness is too hard for either the well or the sick to face. He must hold a strict watch over himself, keeping most of the parts of his body covered, and not chatting with laymen, but saying only what is necessary, since he realizes that perforce this is pertinent to the promotion of healing; and he will never do anything needlessly in these matters, or with showiness. Consider all these questions carefully in order to make your preparations beforehand, so that they can be applied with convenience, since otherwise some difficulty in their application must always result.

8. These practices in medicine must be carried out with great care: palpation, anointing, ablutions, moving the hands with elegance, applying lint, compresses and bandages, and consideration of the effects of weather. In giving medications and treating wounds and conditions of the eye—in all their various forms—you must have at hand instruments, appliances and scalpels etc. you have prepared in advance: to be unprepared in these matters results in helplessness and damage. Also acquire a second,

238 *πρὸς τὰς ἀποδημίας*[14] *ἡ διὰ χειρῶν· ἡ δ' | εὐχερεστάτη διὰ μεθόδων· οὐ γὰρ οἷόν τε περιέρχεσθαι πάντα τὸν ἰητρόν.*

9. *Ἔστω δέ σοι εὐμνημόνευτα φάρμακά τε καὶ δυνάμιες ἁπλαῖ καὶ ἀναγεγραμμέναι—εἴπερ ἄρα ἐστὶν ἐν νόῳ καὶ τὰ περὶ νούσων ἰήσιος—καὶ οἱ τούτων τρόποι, καὶ ὁσαχῶς καὶ ὃν τρόπον περὶ ἑκάστων ἔχουσιν· αὕτη γὰρ ἀρχὴ ἐν ἰητρικῇ καὶ μέσα καὶ τέλος.*

10. *Προκατασκευάσθω δέ σοι καὶ μαλαγμάτων γένεα πρὸς τὰς ἑκάστων χρήσιας, ποτήματα ὑγραίνειν*[15] *δυνάμενα ἐξ ἀναγραφῆς ἐσκευασμένα πρὸς τὰ γένεα. προητοιμάσθω δὲ καὶ τὰ πρὸς φαρμακίην ἐς τὰς καθάρσιας, εἰλημμένα ἀπὸ τόπων τῶν καθηκόντων, ἐσκευασμένα ἐς ὃν δεῖ τρόπον, πρὸς τὰ γένεα καὶ τὰ μεγέθεα ἐς παλαίωσιν μεμελετημένα, τὰ δὲ πρόσφατα ὑπὸ τὸν καιρόν, καὶ τἆλλα κατὰ λόγον.*

11. *Ἐπὴν δ' ἐσίῃς πρὸς τὸν νοσέοντα, τούτων σοι ἀπηρτισμένων, ἵνα μὴ ἀπορῇς, εὐθέτως ἔχων ἕκαστα πρὸς τὸ ποιησόμενον, ἴσθι δὲ γινώσκων ὃ χρὴ ποιέειν πρὶν ἢ ἐσελθεῖν· πολλὰ γὰρ οὐδὲ συλλογισμοῦ, ἀλλὰ βοηθείης δεῖται τῶν πρηγμάτων. προδιαστέλλεσθαι οὖν χρὴ τὸ ἐκβησόμενον ἐκ τῆς ἐμπειρίης· ἔνδοξον γὰρ καὶ εὐμαθές.*

12. *Ἐν δὲ τῇ ἐσόδῳ μεμνῆσθαι καὶ καθέδρης,*

[14] *ἀποδημίας* Linden in text, after Cornarius' *ad peregrinationes*: *ἐπιδημίας* M [15] *ὑγραίνειν* Potter: *τέμνειν* M

more basic set of instruments to carry in your hands on excursions; this is most convenient if methodically arranged, since a physician cannot go though everything.

9. Keep present in your memory medicines and their potencies, both simples and compounds from prescriptions—seeing that the ways to cure diseases reside in the mind—their types and their qualities, and the particular one for each (sc. of the diseases). Indeed, this is the beginning, the middle, and the end of medicine.

10. Make yourself emollients for use in each kind of disease, and drinks able to moisten,[2] prepared by type from prescription. Also provide in advance medicinal cathartics acquired from suitable places, preparing them in the way required, taking account of their kinds and their sizes; preserve some to last a long time, but pluck others fresh at the right moment, and so on in the same way.

11. When you visit a patient, having made the necessary preparations in order to avoid being at a loss, make sure that everything is in good order for what is going to be done, and be clear in your mind before you go in what must be done: many situations will require not reasoning, but some helpful act. So you must predict the disease outcome, on the basis of your past experience, since this will bolster your reputation, and is after all easily learned.

12. When you enter, remember to pay attention to your

[2] The M reading "incise" makes no obvious sense: I am suggesting "moisten" as a possibility.

καὶ καταστολῆς, περιστολῆς, ἀνακυριώσιος, βραχυλογίης, ἀταρακτοποιησίης, προσεδρίης, ἐπιμελείης,
240 *ἀντιλέξιος πρὸς τὰ ἀπαντώμενα, πρὸς τοὺς* | *ὄχλους τοὺς ἐπιγινομένους εὐσταθείης τῆς ἐν ἑωυτῷ, πρὸς τοὺς θορύβους ἐπιπλήξιος, πρὸς τὰς ὑπουργίας ἑτοιμασίης. ἐπὶ τούτοισι μέμνησο παρασκευῆς τῆς πρώτης· εἰ δὲ μή, τὰ κατ' ἄλλα ἀδιάπτωτον, ἐξ ὧν παραγγέλλεται ἐς ἑτοιμασίην.*

13. *Ἐσόδῳ χρέο πυκνῶς, ἐπισκέπτεο ἐπιμελέστερον, τοῖσιν ἀπατωμένοισιν κατὰ τὰς μεταβολὰς ἀπαντῶν· ῥᾷον γὰρ εἴσῃ, ἅμα δὲ καὶ εὐμαρέστερος ἔσῃ. ἄστατα γὰρ τὰ ἐν ὑγροῖσι· διὸ καὶ εὐμεταποίητα ὑπὸ φύσιος καὶ ὑπὸ τύχης· ἀβλεπτηθέντα γὰρ κατὰ τὸν καιρὸν τῆς ὑπουργίης ἔφθασαν ὁρμήσαντα καὶ ἀνελόντα· οὐ γὰρ ἦν τὸ ἐπικουρῆσον. πολλὰ γὰρ ἅμα τὰ ποιέοντά τι χαλεπόν· τὸ*[16] *γὰρ καθ' ἓν κατ' ἐπακολούθησιν εὐθετώτερον καὶ ἐμπειρότερον.*

14. *Ἐπιτηρεῖν δὲ χρὴ καὶ τὰς ἁμαρτίας τῶν καμνόντων, δι' ὧν πολλάκις διεψεύσαντο ἐν τοῖσι προσάρμασι τῶν προσφερομένων· ἐπεὶ*[17] *τὰ μισητὰ ποτήματα* ‹*οὐ*›[18] *λαμβάνοντες, ἢ φαρμακευόμενοι ἢ θεραπευόμενοι, ἀνῃρέθησαν· καὶ αὐτῶν μὲν οὐ πρὸς ὁμολογίην τρέπεται τὸ ποιηθέν, τῷ δ' ἰητρῷ τὴν αἰτίην προσῆψαν.*

15. *Ἐσκέφθαι δὲ χρὴ καὶ τὰ περὶ ἀνακλίσεων, ἃ*

[16] *τὸ* Littré: *τῶν* M [17] *ἐπεὶ* Froben (*siquidem* Calvus): *ἐπὶ* M [18] Add Littré, following Cornarius' *non accipientes*.

manner of sitting, reserved appearance, clothing, authoritative mien, brevity of speech, calming influence, bedside manner, carefulness, and how you reply to objections; project a firm confidence in yourself in the face of troubles that arise, clarify confusions, and show a readiness in rendering services. In these cases remember your first plan; otherwise be steadfast in all the other points included in the instructions for preparedness I have given.

13. Visit patients frequently, paying careful attention during the deceptive periods at all kinds of changes; this way you will understand cases more easily and confidently. The humors are unstable, making them susceptible to the effects of nature and chance: if the correct moment for providing medical assistance goes unnoticed, they may first rush upon a patient and carry him off, since no treatment was given. In cases where many factors are at play simultaneously, there is difficulty, whereas sequences that follow one by one make it easier to recognize the correct cause and collect experience.

14. You must also look for patients' mistakes which often lead them to answer incorrectly about whether or not they are following their prescribed diet: for example, on occasion patients' failing to take the potions they detest, whether purgative or for other therapeutic purposes, has led to their death. However, the errors they have committed never result in an admission of their responsibility, but blame is cast upon the physician.

15. Thought must also be given to where patients' beds

μὲν αὐτῶν πρὸς τὴν ὥρην, ἃ δὲ πρὸς τὰ γένεα· οἱ μὲν γὰρ αὐτῶν ἐς εὐπνόους,[19] *οἱ δ' ἐς καταγείους καὶ σκεπινοὺς τόπους· τά τε ἀπὸ ψόφων καὶ ὀσμῶν, μάλιστα δ' ἀπὸ οἴνου· χειροτέρη γὰρ αὕτη, φυγεῖν δὲ καὶ μετατιθέναι.* |

242 16. *Πρήσσειν δ' ἅπαντα ταῦτα ἡσύχως, εὐσταλέως, μεθ' ὑπουργίης τὰ πολλὰ τὸν νοσέοντα ὑποκρυπτόμενον· ἃ δὲ*[20] *χρή, παρακελεύοντα ἱλαρῶς καὶ εὐδιεινῶς, σφέτερα δὲ ἀποτρεπόμενον, ἅμα μὲν ἐπιπλήσσειν μετὰ πικρίης καὶ ἐντάσεων, ἃ δὲ παραμυθέεσθαι μετ' ἐπιστροφῆς καὶ ὑποδέξιος, μηδὲν ἐπιδείκνυντα τῶν ἐσομένων ἢ ἐνεστώτων αὐτοῖσι. πολλοὶ γὰρ δι' αἰτίην ταύτην ἐφ' ἕτερα ἀπεώσθησαν, διὰ τὴν πρόρρησιν τὴν προειρημένην τῶν ἐνεστώτων ἢ ἐπεσομένων.*

17. *Τῶν δὲ μανθανόντων ἔστω τις ὁ ἐφεστὼς ὅκως τοῖσι παραγγέλμασιν οὐ πικρῶς χρήσηται, ποιήσει δ' ὑπουργίην τὸ προσταχθέν. ἐκλέγεσθαι δ' αὐτῶν ἤδη τοὺς ἐς τὰ τῆς τέχνης εἰλημμένους, προσδοῦναί τι τῶν ἐς τὸ χρέος, ἢ ἀσφαλέως προσενεγκεῖν· ὅκως τε ἐν διαστήμασι μηδὲν λανθάνῃ σε. ἐπιτροπὴν δὲ τοῖσιν ἰδιώτῃσι μηδέποτε διδοὺς περὶ μηδενός· εἰ δὲ μή, τὸ κακῶς πρηχθὲν εἰς σὲ χωρῆσαι τὸν ψόγον ἐᾷ.*[21] *μήποτ' ἀμφιβόλως ἔχῃ, ἐξ ὧν τὸ μεθοδευθὲν χωρήσει, καὶ οὐ σοὶ τὸν ψόγον περιάψει, τευχθὲν δὲ*

[19] *ἐς εὐπνόους* Ermerins, following Foes' n. 37: *ἐς πόνους* M
[20] *ἃ δὲ* Matthiae (Littré): *ἅδε* Zwinger: *ὧδε* M
[21] *χωρῆσαι . . . ἐᾷ* Littré: *χωρήσει τοῦ ψόγου ἐᾶν* M

should be set up, with regard both to the seasons and to the kind of disease. Some (sc. should be placed) in well-aired locations, or in ground-floor rooms or protected areas, while others must be removed from noises or odors, in particular of wine: this is the worst kind of place, and must be vacated and eschewed.

16. All these measures are to be carried out quietly and adroitly, keeping their performance for the most part away from the patient's view; give necessary orders in a quiet and cheerful tone, diverting his attention away from what he is experiencing. On some occasions you must reprove the patient sharply and forcefully, but on others comfort him with solicitude and empathy, revealing nothing of future events or what is threatening him. In fact, in this way many patients have been driven into a desperate state by a pronouncement of the kind I have referred to about what is threatening and imminent.

17. Leave one of yours students in charge to carry out instructions without overexactness, and to provide the help ordered. Let him be selected from among those who are already admitted into the art, so that he is able to furnish whatever is necessary, and he can apply treatment in safety; he should make sure that nothing in the management escapes your notice (sc. while you have been absent). Never give the oversight of anything to a layman, since if you do, it lets blame for the wrong done fall back on you. Also, let there never be any doubt about the principles from which your method of treatment is derived: then no blame will be attached to you, and things that are success-

πρὸς τὸ γάνος[22] ἔσται. πρόλεγε οὖν ταῦτα πάντα ἐπὶ τῶν ποιεομένων, οἷς καὶ τὸ ἐπεγνῶσθαι πρόκειται. |

244 18. Τούτων οὖν ἐόντων τῶν πρὸς εὐδοξίην καὶ εὐσχημοσύνην τῶν ἐν τῇ σοφίῃ καὶ ἰητρικῇ καὶ ἐν τῇσιν ἄλλῃσι τέχνῃσι, χρὴ τὸν ἰητρὸν διειληφότα τὰ μέρεα περὶ ὧν εἰρήκαμεν † περιεννύμενον πάντοτε τὴν ἑτέρην διατηρέοντα φυλάσσειν † καὶ παραδιδόντα ποιέεσθαι· εὐκλεᾶ γὰρ ἐόντα πᾶσιν ἀνθρώποισι διαφυλάσσεται. οἵ τε δι᾽ αὐτῶν ὁδεύσαντες δοξαστοὶ πρὸς γονέων καὶ τέκνων· κἢν τινες αὐτῶν μὴ πολλὰ γινώσκωσιν, ὑπ᾽ αὐτῶν τῶν πρηγμάτων ἐς σύνεσιν καθίστανται.

[22] γάνος Jones: γένος M

ful will be a source of pride. Thus it is good to predict all these things while they are unfolding to your assistants for whom a fuller knowledge is appropriate.

18. Since these points are capital for establishing a good reputation and dignity in both wisdom and medicine (as well as in the other arts), the physician must mark off the parts I have discussed—enclosing one kind (sc. of wisdom) permanently, and keeping it carefully guarded—and pass them on to be performed: for these, being highly valued, are treasured by all people. Physicians who walk in this path are held in honor by their parents and their children, and even if some of them do not know that many things, by carrying out the procedures themselves they will advance in their understanding.

DENTITION

INTRODUCTION

Dentition is a short collection of prognostic aphorisms pertaining to childhood conditions of the throat and mouth, devoted to two kinds of conditions, in chapters as follows:

1–17, 19, 28–29	Formation of the teeth, with remarks on the nursing and weaning of infants.
18, 20–27, 30–32	Ulceration of the tonsils, uvula, and throat.

None of these aphorisms is present in the main Hippocratic aphoristic books (*Aphorisms*, *Prorrhetic I*, *Coan Prenotions*), although a few ideas are shared. Jones' study of the treatise's vocabulary (p. 319f.) led him to suggest a date after 400 BC.

The work is preserved in only one independent Greek manuscript (V) and its copies, and there is no reference to it in medieval times. No special study has been devoted to the writing, but it is contained in the collected editions, including Littré and Ermerins, and in:

Joly, R. *Hippocrate. . . . De la Dentition*. Budé XIII. Paris, 1978.

The present edition is based completely on the work of my predecessors.

ΠΕΡΙ ΟΔΟΝΤΟΦΥΙΗΣ

VIII 544 Littré

1. Τὰ φύσει εὔτροφα τῶν παιδίων οὐκ ἀνάλογον τῆς σαρκώσεως καὶ τὸ γάλα θηλάζει.

2. Τὰ βορὰ καὶ πολὺ ἕλκοντα γάλα οὐ πρὸς λόγον σαρκοῦται.

3. Τὰ πολὺ διουρέοντα τῶν θηλαζόντων ἥκιστα ἐπιναύσια.

4. Οἷσι πολλὴ φέρεται ἡ κοιλίη καὶ εὐπεπτοῦσιν, ὑγιεινότερα· ὁπόσοισιν ὀλίγη, βοροῖσιν ἐοῦσι καὶ μὴ ἀνάλογον τρεφομένοισιν, ἐπίνοσα.

5. Ὁπόσοισι δὲ πολὺ γαλακτῶδες ἀπεμεῖται, κοιλίη ξυνίσταται.

6. Ὁπόσοισιν ἐν ὀδοντοφυΐῃ ἡ κοιλίη πλείω ὑπάγει ἧσσον σπᾶται ἢ ὅτῳ ὀλιγάκις.

7. Ὁπόσοισιν ἐπὶ ὀδοντοφυΐῃ πυρετὸς ὀξὺς ἐπιγίνεται ὀλιγάκις σπῶνται.

8. Ὁπόσα ὀδοντοφυεῦντα εὔτροφα μένει καταφορικὰ ἐόντα κίνδυνος σπασμὸν[1] ἐπιλαβεῖν.

9. Τὰ ἐν χειμῶνι ὀδοντοφυεῦντα, τῶν ἄλλων ὁμοίων ἐόντων, βέλτιον ἀπαλλάσσει.

[1] -μον Aldina: -μος V

DENTITION

1. Well-nourished children do not suck milk in proportion to their corpulence.

2. Those who are voracious and draw much milk do not gain flesh in proportion.

3. Those at the breast who pass much urine are the least subject to vomiting.

4. Those whose cavity passes copious stools and who digest well are healthier; those who pass little, and although voraciously hungry are not nourished in proportion, are sickly.

5. Those who vomit up much milky material are constipated in their cavity.

6. Those whose cavity excretes downward more often during teething have less seizures than those who excrete less often.

7. Those who at the time of teething have an acute fever rarely experience seizures.

8. Those who during teething remain well-nourished although they are lethargic are in danger of having a seizure.

9. Those who teethe in winter, everything else being the same, recover better.

10. Οὐ πάντα τὰ ἐπὶ ὀδοῦσι σπασθέντα τελευτᾷ· πολλὰ δὲ καὶ διασῴζεται.

11. Τὰ μετὰ βηχὸς ὀδοντοφυεῦντα χρονίζει·[2] ἐν δὲ τῇ διακεντήσει ἰσχναίνεται μᾶλλον.

12. Ὁπόσα ἐν τῷ ὀδοντοφυεῖν χειμῶνας ἔχει, ταῦτα καὶ προσεχόντως ἠγμένα ῥᾷον φέρει ὀδοντοφυΐαν.

13. Τὰ διουρεῦντα πλέον ἢ διαχωρεῦντα πρὸς λόγον εὐτροφώτερα. |

546 14. Ὁπόσοισιν οὐρεῖται μὴ πρὸς λόγον, κοιλίη δὲ πυκνῶς ὠμὸν ἐκ παιδίων παρηθεῖ, ἐπίνοσα.

15. Τὰ εὔυπνα καὶ εὔτροφα πολὺ ἀναλαμβάνειν παράκειται καὶ[3] οὐχ ἱκανῶς διῳκημένον.

16. Τὰ παρεσθίοντα ἐν τῷ θηλάζειν ῥᾷον φέρει ἀπογαλακτισμόν.

17. Τὰ πολλάκις παρηθεῦντα[4] δίαιμον καὶ ἄπεπτον κατὰ κοιλίην πλεῖστα [τῶν][5] ἐν πυρετῷ ὑπνώδεα.

18. Τὰ ἐν παρισθμίοις ἕλκεα ἄνευ πυρετῶν γινόμενα ἀσφαλέστερα.

19. Ὁπόσοισιν ἐν τῷ θηλάζειν τῶν νηπίων βὴξ προσίσταται, σταφυλὴν εἴωθε μείζονα ἔχειν.

20. Ὁπόσοισι ταχέως ἐν παρισθμίοις νομαὶ ἐφίστανται, τῶν πυρετῶν μενόντων καὶ βηχίων, κίνδυνος πάλιν γενέσθαι ἕλκεα.

[2] -νίζει Aldina: -νίζειν V

[3] παράκειται καὶ Joly: καὶ παράκειται V

[4] παρηθ. Foes (n. 4): παριθ. V

[5] Del. Joly

10. Not all that have seizures while they are teething die, but many also recover.

11. Those that have a cough while they are teething continue longer; while the teeth are erupting, they become more emaciated.

12. Those that have a tempestuous time during teething, if this is carefully managed, will tolerate the teething more easily.

13. Those that pass more urine in proportion to their stools are better nourished.

14. Those who do not pass a proportionate amount of urine, but whose cavity frequently excretes undigested stools from infancy on, are sickly.

15. Those who sleep well and are well-nourished are able to ingest much, and even if it is not properly prepared.

16. Those who eat food along with being nursed tolerate weaning more easily.

17. Those who often excrete bloody and undigested stools through their cavity are more often drowsy in a fever.

18. Lesions on the tonsils when unaccompanied by fever are less dangerous.

19. Those infants attacked by a cough while they are nursing normally have a swollen uvula.

20. Those who quickly develop corroding lesions on their tonsils, if the fevers and mild coughs persist, are in danger that the lesions will recur.

21. Τὰ παλινδρομήσαντα ἐν ἰσθμίοις ἕλκεα τοῖσι νηπίοισι[6] κινδυνώδεα.

22. Τοῖσι παιδίοισιν ἀξιολόγοις ἕλκεσιν ἐν παρισθμίοισι, καταπινομένων, σωτηρίας ἐστίν, ὁπόσα δὴ μᾶλλον τῶν πρότερον μὴ δυναμένων καταπίνειν.

23. Ἐν <τοῖσιν ἐν>[7] παρισθμίοις ἕλκεσι, πολὺ τὸ χολῶδες ἀνεμεῖσθαι ἢ κατὰ κοιλίην ἔρχεσθαι, κινδυνῶδες. |

548 24. Ἐν τοῖσιν ἐν παρισθμίοισιν ἕλκεσιν ἀραχνιῶδές τι ἐὸν οὐκ ἀγαθόν.

25. Ἐν τοῖσιν ἐν παρισθμίοισιν ἕλκεσι μετὰ τοὺς πρώτους χρόνους διαρρεῖν φλέγμα διὰ τοῦ στόματος, πρότερον οὐκ ὄν, χρήσιμον, ὅμως ἀνακτέον· ἢν δὲ ἄρξηται ξυνδιδόναι, πάντως ἀσμενιστέον· τὸ δὲ μὴ οὕτως διαρρέον εὐλαβητέον.

26. Ῥευματιζομένοις παρίσθμια κοιλίη κατενεχθεῖσα πλείω λύει τὰς ξηρὰς βῆχας· παιδίοισιν ἀνενεχθέν τι πεπεμμένον πλείω λύει.

27. Τὰ πολὺν χρόνον ἐν παρισθμίοις ἕλκεα ἀναυξῆ μένοντα ἀκίνδυνα πρὸ τῶν πέντε ἢ ἓξ ἡμερέων.

28. Τὰ πολὺ γάλα τῶν θηλαζόντων ἀναλαμβάνοντα ὡς τὸ πολὺ ὑπνώδεα.

[6] νηπίοισι Ermerins after Cornarius' *in infantibus*: ὁμοίοισι V

[7] Add. Joly

21. Recurrent lesions on the tonsils bring danger to infants.

22. For serious lesions on the tonsils in small children who can nevertheless still drink, there is a chance of cure, and all the more so if they were not able to drink before.

23. In lesions on the tonsils, for bilious material to be vomited up or to be excreted by way of the cavity is a sign of danger.

24. In lesions on the tonsils the presence of material like cobwebs is not a good sign.

25. In lesions on the tonsils, if after the first while phlegm runs out through the mouth, when it did not do so before, this is a valuable sign, but still it must be brought up.[1] If the flow begins to abate, this is very welcome; but if the flow does not take place like this, attention must be paid.

26. When the tonsils are having fluxes, downward excretion of the cavity tends more to relieve dry coughs; in children, however, it is usually an upward movement of concocted material that brings more relief.

27. Lesions on the tonsils that persist for a long time[2] without increasing are not a sign of danger before the fifth or sixth day.

28. Nursing children that consume much milk are generally sleepy.

[1] This is Jones' interpretation; Joly translates: "pourtant, il faut le faire cesser."

[2] Littré and Jones question the apparent contradiction between "for a long time" and "the fifth or sixth day." Jones suggests that the former is a gloss on the latter.

29. Τὰ μὴ εὔτροφα[8] τῶν θηλαζόντων ἄτροφα καὶ δυσανάληπτα.

30. Ἕλκεα ἐν θέρει γιγνόμενα ἐν παρισθμίοισι χείρονα τῶν ἐν τῇσιν ἄλλῃσιν ὥρῃσι· τάχιον γὰρ νέμεται.

31. Τὰ περὶ σταφυλὴν νεμόμενα ἕλκεα ἐν παρισθμίοισιν, σῳζομένοισι τὴν φωνὴν ἀλλοιοῖ.

32. Τὰ περὶ φάρυγγα νεμόμενα ἕλκεα χαλεπώτερα καὶ ὀξύτερα· ὡς ἐπὶ πολὺ δύσπνοιαν ἐπιφέρει.

[8] εὔτροφα Ermerins: εὐτροφέα V

29. Nursing children that are not being well nourished lose flesh and have difficulty recovering it.

30. Lesions that arise on the tonsils in summer are worse than those in the other seasons, since they tend to spread more rapidly.

31. Lesions on the tonsils that spread around the uvula change the voice in patients who recover.

32. Lesions spreading through the throat are very serious and acute—they generally bring on difficult breathing.

INDEX